Conservation and Biodiversity Banking

The Ecosystem Marketplace seeks to become the world's leading source of information on markets and payment schemes for ecosystem services (services such as water quality, carbon sequestration and biodiversity). We believe that by providing reliable information on prices, regulation, science and other market-relevant factors, markets for ecosystem services will one day become a fundamental part of our economic system, helping give value to environmental services that, for too long, have been taken for granted. In providing useful market information, we hope not only to facilitate transactions, but also to catalyse new thinking, spur the development of new markets, and achieve effective and equitable nature conservation. The Ecosystems Marketplace is a project of Forest Trends.

www.ecosystemmarketplace.com

List of Figures, Tables and Boxes

Figures

Tables

Boxes

Conservation and Biodiversity Banking

A Guide to Setting Up and Running Biodiversity Credit Trading Systems

Edited by Nathaniel Carroll, Jessica Fox and Ricardo Bayon

London • Sterling, VA

First published by Earthscan in the UK and USA in 2008

ISBN-13: 978-1-84407-471-6
Typeset by Safehouse Creative
Printed and bound in the UK by TJ International Ltd
Cover design by Yvonne Booth

For a full list of publications please contact:

Earthscan
8–12 Camden High Street
London, NW1 0JH, UK
Tel: +44 (0)20 7387 8558
Fax: +44 (0)20 7387 8998
Email: earthinfo@earthscan.co.uk
Web: **www.earthscan.co.uk**

22883 Quicksilver Drive, Sterling, VA 20166-2012, USA

Earthscan publishes in association with the International Institute for Environment and Development

A catalogue record for this book is available from the British Library

Library of Congress Cataloging-in-Publication Data

Conservation and biodiversity banking : a guide to setting up and running biodiversity credit trading systems / edited by Nathaniel Carroll, Jessica Fox and Ricardo Bayon.
p. cm.
ISBN-13: 978-1-84407-471-6 (hardback)
ISBN-10: 1-84407-471-4 (hardback)
1. Biodiversity conservation – Economic aspects. 2. Environmental impact charges. 3. Environmental responsibility – Economic aspects. I. Bayon, Ricardo. II. Fox, Jessica. III. Carroll, Nathaniel.
QH75.C655 2007
333.95'16 –dc22

2007034791

The paper used for this book is FSC-certified and totally chlorine-free. FSC (the Forest Stewardship Council) is an international network to promote responsible management of the world's forests.

Contents

Forewords

I

This book represents a step in a journey of convergent thinking between the business and conservation communities. Five years ago it would have been unthinkable for the CEO of a major mining company to write this foreword, encouraging people to read, evaluate and test the ideas this book sets out. In those five years both communities have engaged and evolved new ways to see our roles now and in the future.

Following recognition that traditional approaches to conservation are not adequate to meet our global biodiversity goals, many in the conservation community now believe that the private sector's power to change outcomes can be influenced to deliver conservation gains.

Rio Tinto has set itself the goal to have a net positive impact on biodiversity; this means that our actions have positive effects on biodiversity features and their values that outweigh the inevitable negative impacts of our activities. Biodiversity offsets are a critical element in achieving our goal. Voluntary markets for biodiversity present an opportunity for Rio Tinto to capitalize on its extensive land holding, potentially enabling operations to trade credits they generate through biodiversity protection and enhancement. This book is a major step forward in creating a framework for comparing biodiversity value, and we welcome it.

As with all innovative initiatives there is still much to learn and improve. It is also clear that voluntary markets and government legislation are already taking the subject from ideas to actions. The book provides guidance to those who see the opportunity to test ideas of investment, trading and financial return in biodiversity work. Only by testing these ideas in real life will we find out how to make them work well and where the boundaries for their use may lie.

It would be easy to portray this book as conservation packaged in a way that business can understand and engage with, and a clearer link between today's actions and tomorrow's value will certainly encourage business actions that favour conservation. That is only part of the benefit we should look for; all sectors of society need as many mechanisms that reduce biodiversity loss as possible, and the guidance contained in the book promotes the use of market mechanisms by all.

Whatever your affiliation, I encourage you to read this book and to use it as a basis for talking with others active in conservation; its message is that market mechanisms can deliver outcomes that matter to us all.

Tom Albanese
Chief Executive Officer
Rio Tinto
October 2007

II

Around the world species are disappearing and natural systems are being degraded at an alarming rate. According to the Millennium Ecosystem Assessment, an extensive assessment of the state of the world's ecosystems published in 2005, 10 to 30 per cent of mammal, bird and amphibian species are currently threatened with extinction as a result of human actions, and the services that nature provides to human beings are in serious decline.

We have not yet addressed these concerns with adequate focus and resources. One important reason for the decline in ecosystems and species is that markets and policies tend not to value biodiversity and natural systems. There are few rewards for conserving biodiversity and often no penalties for destroying it. Conservation banking and biodiversity offsets provide a new awareness and innovative tools to address this gap.

Conservation banking is an important emerging mechanism for conserving biodiversity. It provides an opportunity for protecting key areas of high biological diversity and areas where endangered species are threatened. Species mitigation banking is an example of a market that increases people's value of biodiversity. Likewise, biodiversity offsets provide a mechanism for ensuring that unavoidable adverse impacts to biodiversity from development are counterbalanced by gains, with the overall aim of achieving a net beneficial outcome for conservation. While more testing is needed, these tools represent creative new approaches for managing environmental liabilities associated with development in a positive manner for conservation.

For over 50 years, The Nature Conservancy has worked to conserve species and natural systems in forests, grasslands, deserts, rivers and oceans. Our commitment is to the diversity of life on Earth. Our approach has been to promote conservation-friendly public policies and private land conservation, including land acquisition, and to enable local communities, governments and investors to make informed choices about where investments in conservation are most effective. Increasingly, we recognize the importance of incentive-based mechanisms for conserving nature.

We have experienced success with a variety of innovative conservation and finance mechanisms, and are now testing the effectiveness of market-based mechanisms and payments for ecosystem services in an array of settings, as tools for conserving nature.

As conservation banking and biodiversity offsets move from concept to practice, it is critical to test these tools widely, develop appropriate principles and standards for their implementation and share lessons from experiences to determine their effectiveness for conservation. *Conservation and Biodiversity Banking* by Nathaniel Carroll, Jessica Fox and Ricardo Bayon is a timely and valuable foundation for moving this process forward. Written for a wide audience it provides a comprehensive and practical guide for understanding how species credit banking works, and how practitioners and those interested in establishing a species bank might proceed.

The book comes at a time when current trends show rapid economic growth in middle income countries like China and India and rising investment in development worldwide, particularly in resource extraction and infrastructure projects. We need to look to what the future holds for conservation and prepare for it with the right set of tools. Given projections for unprecedented development, tools like conservation banking and biodiversity offsets that support sustainable development will help all of us, working together, achieve better conservation outcomes and ensure a better future for nature and humankind.

Stephanie Meeks
Acting President and Chief Executive Officer
The Nature Conservancy
October 2007

List of Authors

Tundi Agardy is internationally renowned in marine conservation, with extensive field and policy experience in African, Asian, Caribbean, Mediterranean and North American regions. She currently heads Sound Seas, an independent policy group specializing in coastal planning and assessment, marine protected areas, fisheries management and market-based conservation. Formerly global marine program senior director at Conservation International and senior scientist for WWF, she has assisted non-governmental organizations (NGOs), government agencies and multilateral organizations in conservation planning, projects and programme evaluation. She received her PhD and MMA from the University of Rhode Island, was postdoctoral fellow at Woods Hole Oceanographic Institution, and attended Wellesley and Dartmouth Colleges.

Ricardo Bayon is the director of the Ecosystem Marketplace. For nearly a decade he has been focusing on issues related to finance, socially responsible investment (SRI), and the environment. He has been a fellow of the New America Foundation and done work for a number of organizations, including Innovest Strategic Value Advisors, Domini Social Investments, the International Finance Corporation (IFC) of the World Bank, Forest Trends, The Nature Conservancy, the UN Foundation, the World Conservation Union (IUCN) and the Inter-American Development Bank, among others.

Robert Bonnie is co-director of the Land, Water and Wildlife Program at Environmental Defense. He is also managing director of the Center for Conservation Incentives there. Bonnie joined Environmental Defense in 1995 and has focused on conservation of wildlife and biodiversity on private lands through economic incentives. He has been extensively involved in the development of the Safe Harbor program, which has resulted in millions of acres of privately owned land being managed cooperatively for endangered species. Bonnie also has extensive experience in reform and implementation of Farm Bill incentive programmes. He is a leading expert on the use of markets as a means to reward stewardship on farms, ranches and forest lands, including through endangered species conservation banking and carbon sequestration crediting on forest and farm lands. Bonnie has an AB from Harvard College and a Masters in Environmental Management and Forestry from Duke University's Nicholas School of the Environment.

Howard Brown has a BSc in fishery biology from Humboldt State University in Arcata, California. Howard has been working in the field of salmon and steelhead monitoring, protection and restoration in California for over 15 years, and is currently employed by National Oceanic and Atmospheric Administration's (NOAA) National Marine Fisheries Service Sacramento Office of Protected Resources, where he works with federal, state and local entities on Endangered Species Act consultations for flood control, fish screens, fish habitat restoration and hydroelectric projects.

Tom Cannon is an aquatic ecologist with degrees in fisheries and biostatistics. He serves as the program manager for Wildlands aquatic programmes, coordinating business development and project entitlement in Wildlands' five regions of operation in the West. He has over 30 years' experience in the environmental consulting field, where his specialties have been project management, permitting, project planning and implementation, river and estuary ecology, habitat restoration, monitoring, mitigation and environmental assessment. He has worked with Wildlands for over five years, including bank development and monitoring. Most recently, Tom has worked with the Federal Services in the development of conservation banks for listed fish.

Nathaniel Carroll is project manager for the Ecosystem Marketplace. Before joining Forest Trends, Nathaniel worked as a consultant for a private forestry and real estate company in Panama, channelling private investment to restore degraded lands and generate profit from native species forestry. He spent two years with Conservation International's Center for Applied Biodiversity Science, one with the Rapid Assessment Program and the other with the Conservation Tools Program. He has over three years' experience conducting ecological research, from the Rocky Mountains to the Andes, from the Northwest Hawaiian Islands to the Penobscot Bay. Nathaniel holds a BSc from Tufts University and a Masters in Forest Science from Yale University.

Craig Denisoff is vice president and founding partner of Westervelt Ecological Services, a national environmental mitigation company and subsidiary of The Westervelt Company – a 122-year-old natural resources company. Craig also served as president of the National Mitigation Banking Association. He has worked for the US Army Corps of Engineers, US Environmental Protection Agency, California State Legislature, and was assistant secretary for the California Resources Agency supervising the States' national recognized wetland and coastal programmes. Craig has consulted on and developed mitigation policies for California, the US and international governments.

Deborah Fleischer is founder and principal of Green Impact (www.greenimpact.com). She works with land trusts and land conservation organizations that are struggling to identify innovative strategies for protecting and restoring sensitive habitats. Deborah has written for the Ecosystem Marketplace on the pros and cons of wetland mitigation banking, and conducted stakeholder interviews to assess the success of past wetland mitigation projects in the San Francisco Bay Area for National Audubon. She has a Masters in Environmental Studies from Yale University and a Masters in Public Administration from Harvard University.

Jessica Fox holds a BSc from the University of California at Davis and an MSc from Stanford University in Biological Sciences, and is certified by the Ecological Society of America. She leads the Electric Power Research Institute's (EPRI) research in natural resource credit markets, focusing on conservation banking and water quality trading. She has conducted benchmark research in conservation banking, generating the first database of species credit transactions in the US. For the past five years, she has been practically applying her academic research to generate a business-based argument for conservation. Her work has been printed by the *Journal of Conservation Biology*, Environmental Law Institute, Island Press, Edison Electric Institute and others.

Royal C. Gardner is professor of law and director of the Institute for Biodiversity Law and Policy at Stetson University College of Law. His teaching and scholarship focus on market-based approaches to protecting the environment. In 1999–2001, Professor Gardner was appointed to the National Research Council's Committee on Mitigating Wetland Losses. He is currently the chair of the US National Ramsar Committee, the North American representative to the Ramsar Scientific and Technical Review Panel, and a vice chair of the American Bar Association's International Environmental Law Committee. In 2006, he received the National Wetlands Award for Education and Outreach.

Susan Hill is a senior biologist for the US Fish and Wildlife Service, working in the Sacramento Fish and Wildlife Office since 2001. She serves as the Conservation Bank coordinator, administering and developing the conservation bank programme through coordination with a variety of federal, state and local agencies, and public and private sponsors. Prior to working for the Service, Susan worked for the US Army Corps of Engineers in the Planning Division, primarily assisting in emergency levee repairs, designing biological remediation for levee repair sites and writing Environmental Assessments. She has a Bachelor of Science degree in Environmental Science with a minor in biology, and specializes in riparian restoration. She also has a Juris Doctor degree from UOP McGeorge School of Law and is an active, licensed attorney in California.

Mira Inbar is currently pursuing her MBA at the Haas School of Business of the University of California at Berkeley. Prior to joining Haas, Mira worked with Forest Trends, where she designed, cultivated and managed the Business and Biodiversity Offset Program, and coordinated the activities of the international Katoomba Group. At Forest Trends, Mira authored several articles and primers on Payments for Ecosystem Services. Mira graduated from Oberlin College with a biology degree. Shortly thereafter she volunteered with Quechuan communities in the Urubamba River Valley, Peru where she initiated a community reforestation and 'cleaner burning stove' programme. She has also worked with Environmental Defense and the Environmental Resources Trust.

Deborah L. Mead has a BSc in Zoology from the University of California at Davis and an MA in Ecology and Systematic Biology from San Francisco State University. She is currently a biologist in the Endangered Species Program of the US Fish and Wildlife Service (USFWS) in Arlington, Virginia, where she works on Endangered Species Act issues and serves as the national conservation banking coordinator. Past positions include three years as a biologist for the US Geological Survey, two years as a biological consultant for a private firm, two years teaching community college courses in entomology and general biology, and 12 years as an entomologist and biologist for the US Fish and Wildlife Service. She developed the conservation banking programme for the USFWS Sacramento Fish and Wildlife Office, and worked with many private landowners to establish conservation banks for federally listed species in California.

James Shields has spent most of his 30-year career as a wildlife manager and researcher in Australia. Working first for the Australian Museum in Sydney, and then the New South Wales Forestry Commission, Dr Shields has published two books on Australian birds, and conducted research into the effects of forestry operations on plant communities, birds, mammals and invertebrates. A practical background in agriculture (he grew up on a farm in Kansas) and the need to balance conservation costs with forestry profits led to his interest and work on biodiversity banking. Jim lives on the far south coast of New South Wales at present.

Kerry ten Kate is the director of the Business and Biodiversity Offsets Program (Forest Trends). Formerly a barrister in London, Kerry served on the Secretariat of the United Nations Conference on Environment and Development (the 'Rio Earth Summit') before advising governments, companies, investors and the UN on conservation and sustainable development strategies, on which she has written and broadcast extensively. She was policy adviser at the Royal Botanic Gardens, Kew and a member of the UK delegation to negotiations at the UN's Convention on Biological Diversity and Food and Agriculture Organization. For four years, she was director of Investor Responsibility at Insight Investment and is a member of the UK Government's Darwin Advisory Committee.

Sherry Teresa is founder and executive director of the Center for Natural Lands Management (CNLM). Prior to that she worked for the California Department of Fish and Game and the USFWS. Since the creation of the Center, Sherry has developed the habitat management programme, the Property Analysis Record (PAR) and other tools to assist in the perpetual stewardship of mitigation and conservation lands, as well as authoring numerous articles on conservation stewardship. One significant aspect of the Center's mission is to develop, promote and disseminate the science of conservation land management. The CNLM currently has over 52,000 acres under management, including 17 mitigation and/or conservation banks, and consults nationwide on habitat stewardship and conservation planning. Sherry completed a Bachelors degree in Zoology at Brigham Young University and a Masters degree in Biogeography and Ecosystems Analysis at UCLA.

Wayne White spent 31 years with the USFWS. As the chief of Endangered Species Program and a field supervisor in Region 1 of USFWS, Wayne developed Endangered Species Act policy in the early years of implementing the Act. While managing the USFWS's largest field office in Sacramento, California, he became very involved with developing policy for conservation and mitigation banks for USFWS. Wayne now runs his own consulting business, W-Squared Consulting, and works with mitigation/conservation banking companies helping them to investigate new markets around the country. He is also working with the Business and Biodiversity Offset Program to help corporations assess their biodiversity impacts and design offsets, and developing a handbook for corporations and governments to use in the development of policies for offsets.

David S. Wilcove is professor of ecology, evolutionary biology and public affairs at Princeton University. Prior to joining Princeton's faculty, he worked for Environmental Defense, The Wilderness Society and The Nature Conservancy. He is the author of *No Way Home: The Decline of the World's Great Animal Migrations* (2007), *The Condor's Shadow: The Loss and Recovery of Wildlife in America* (1999), and numerous technical and popular articles in the fields of conservation biology, ornithology and wildlife conservation. He has served on the boards of directors of the Society for Conservation Biology, Rare, American Bird Conservancy, Natural Areas Association and the New Jersey Audubon Society.

Acknowledgements

We are tremendously indebted to all the authors for the time, effort and knowledge they have generously shared with us. We would also like to thank the members of the Katoomba Group for their support.

List of Acronyms and Abbreviations

BBOP	Business and Biodiversity Offset Program
BEI	bank enabling instrument
CARR	Comprehensive, Adequate and Representative Reserve
CBA	conservation bank agreement
CBD	Convention on Biological Diversity
CBRT	conservation bank review team
CDFG	California Department of Fish and Game
CEQA	California Environmental Quality Act
CERCLA	Comprehensive Environmental Response, Compensation and Liability Act
CERES	California Environmental Resources Evaluation Center
CESA	California Endangered Species Act
CMA	Catchment Management Authority
CNLM	Center for Natural Lands Management
CPI	Consumer Price Index
CWA	Clean Water Act
DECC	Department of Environment and Climate Change
DPS	distinct population segment
EEC	Endangered Ecological Communities
EFH	essential fish habitat
EIA	environmental impact assessment
ELI	Environmental Law Institute
EPA	Environmental Protection Agency
EPRI	Electric Power Research Institute
ESA	Endangered Species Act
ESU	ecologically significant unit
GAAP	generally accepted accounting principles
GSEE	governance, social, environmental and ethical
HCP	habitat conservation plan
I&C	initial and capital (period)

IFC	International Finance Corporation
IRS	Internal Revenue Service
ITP	incidental take permit
ITS	incidental take statement
IUCN	World Conservation Union
KTP	key threatening processes
LLC	limited liability company
MA	Millennium Ecosystem Assessment
M&M	monitoring and maintenance (period)
MAWSS	Mobile Area Water and Sewer System
MBRT	mitigation bank review team
MOA	memorandum of agreement
MOU	memorandum of understanding
MSA	Magnuson-Stevens Fishery Conservation Act
MSCP	multiple species conservation plan
NCCP	natural community conservation plan
NEPA	National Environmental Policy Act
NFP	National Forest Policy
NGO	non-governmental organization
NMFS	National Marine Fisheries Service
NOAA	National Oceanic and Atmospheric Administration
NRCS	Natural Resource Conservation Service
NSW	New South Wales
NVA	Native Vegetation Act
ODFW	Oregon Department of Fish and Wildlife
ODOT	Oregon Department of Transportation
PAR	Property Analysis Record
RFA	Regional Forest Agreement
RIBITS	Regional Internet Bank Information Tracking System
SEC	Security and Exchanges Commission
SFWO	Sacramento Fish and Wildlife Office
SITLA	State Institutional Trust Lands Administration
SRBPP	Sacramento Riverbank Bank Protection Project
SRI	socially responsible investment
TAP	threat abatement plan
TET	The Environmental Trust
TSCA	Threatened Species Conservation Act
TVA	Tennessee Valley Authority
UBA	umbrella bank agreement
UMIFA	Uniform Management of Institutional Funds Act
UPD	Utah prairie dog
USACE	United States Army Corps of Engineers
USFWS	United States Fish and Wildlife Service
USGAO	United States General Accounting Office
VELB	valley elderberry longhorn beetle
VSP	viable salmonid population

Part I

Overview

1

Introduction

Ricardo Bayon, Nathaniel Carroll and Jessica Fox

Species go extinct practically every day and we lose yet another ecosystem to human development. The problem is not new, it has been going on since time immemorial: humans and wildlife have always competed for food, resources and land. Some forms of biodiversity are seen as valuable; others as expendable. The valuable plants and animals have usually been those that provided us with food, shelter, clothing and other such useful products. The rest don't even make it onto our balance sheets. Of course, we have enjoyed wildlife protected in parks – and have recognized that some small species are important for cultural and spiritual reasons. But, for the most part, we have not seen the value in many aspects of biodiversity, and this means that in a head-on collision between biodiversity and a golf course (or some other form of human use), the golf course will win. Rainforests recede in favour of palm oil plantations, mangroves make way for shrimp farms and wetlands wind up losing to Wal-Marts.

The issue may not be new, but the problem is getting worse. As human populations expand, as our global 'footprint' gets bigger, our impact on species and ecosystems increases and the amount of biodiversity left on the planet gets smaller. It is a classic economic case of demand far outstripping supply. And it has now reached a point where we are, quite literally, wiping entire species off the map.

We have known about the problem of dwindling biodiversity for decades. Documentary after documentary, environmental campaign after environmental campaign and scientist after scientist, have all warned us about the issues surrounding our impact on the environment and we feel a deepening discomfort and sense of foreboding.

As is usually the case, our intuition is right. There is something wrong. Healthy biodiversity makes substantial contributions to the functioning of ecosystems, and provides us with numerous services on which our survival depends; services such as the decomposition of waste, pest control, climate regulation, nutrient cycling, oxygen generation and genetic diversity contributing to medicines, to name but a few. In our demands for food, shelter, golf courses, large homes, shopping malls and local airports we disregard species, habitat and the ability of the remaining ecosystems to function.

We delude ourselves into believing that it is not our fault, that blame belongs to some poor farmer in the Amazon, or to evil corporations intent on reaping yet another million in profits. But the truth is that part of the problem resides in us, in our economic system, in the way we give greater value to forests once they are burned down and turned into soy fields, than we do when they are left standing. It is a problem of both values *and* value. When we value iPods over whale pods, our economic system will deliver ever-more species of iPods and wipe out yet another species of whales. When wetlands are seen as nothing more than mosquito-infested swamps, they will give way to shopping malls. And as land becomes ever-more scarce, the problem will simply be aggravated.

We tell ourselves that the system must be broken, but in truth it is doing exactly what it was set up to do: deliver more of what we want and less of what we don't put a value on.

It is no mistake that the suburban landscape in the US appears to be dotted with many more golf courses than wetlands. It is simply a symptom of an economic system that has its values – used here in the sense of its prices – wrong. It is what economists, in their jargon, call a problem of externalities. Some values – like the value of a species of woodpecker, or the value of a particular ecosystem like a rainforest or a wetland – do not enter into the economic system; they are external to it and, as such, aren't taken into account when economic decisions are made.

We like to say that nature is priceless, so valuable that you can't even possibly assign it a dollar value. But it isn't true. The truth is that for eons the price of nature has been woefully close to zero. Supply has outstripped demand to such an extent that we confuse priceless with worthless. But that equation is changing. Priceless nature is becoming economically valuable, in some cases more valuable than a golf course.

That, at its very core, is what this book is about. It is about the use of one particular tool – conservation banking or species banking – as a way of giving species and ecosystems a seat at the economic decision-making table. It covers, in some considerable depth, this one approach being taken (particularly in the US) to give financial value to healthy biodiversity. Conservation banking, species banking or biodiversity banking goes by many names, but in essence it involves government setting a limit on the harm that may come to species, and then allowing the market to resolve the cost of offsetting impacts above the limit or 'cap'. The market, operating under traditional forces of supply and demand, then determines the financial value of the biodiversity. In some cases, as in the US, the government also requires land developers to offset the damage they do to biodiversity, thereby mandating demand for biodiversity protection.

Of course, the actual workings of the system are much more complicated than that. In the US they involve mitigation bank review teams (or MBRTs), various government agencies, biological opinions, lawyers and many, many other layers of necessary complication – and you can read about them in detail in this book – in order to ensure that the laws are truly being adhered to. Just like in the global markets for stocks and bonds we still need a Securities and Exchanges Commission (SEC) to keep things honest. The point is, however, that stripped of these complexities, the outcome of species banking is that damaging species and ecosystems costs more, while conserving species and ecosystems becomes profitable.

Does it work? Well, it is probably too early to tell.

There are encouraging signs across the US that this system, and its sister system of wetland banking, are indeed having the desired impact – making development on sensitive lands more costly and helping create a whole new generation of entrepreneurs whose main and only business is the conservation of biodiversity. But the reality is that we haven't really done the necessary studies, and we haven't had the necessary time to know whether the system is slowing down our society's historic destruction of species and ecosystems. What we can say is that, to paraphrase Winston Churchill's famous quip, it may be the worst of all possible systems except for all the others. Our hope is that this book will help you better understand the system's utility – and perhaps even help you to shape its future.

As far as we're concerned, however, the system has merit. Theoretically and conceptually, the idea that healthy, living biodiversity should be considered economically valuable makes sense. There are clear advantages (Chapter 3) and, of course, pitfalls (Chapter 4). The system could be made better of course, but at least we think it is headed in the right direction.

To illustrate this point let us give you two examples: one hypothetical and one very real.

First, imagine that you are a landowner in the US, somewhere in the south east. You've just inherited land from your parents, land that has been in the family for generations. You love the land, but are also cognisant that you must make a living and that the property taxes and maintenance on the property are beginning to be a drain on your resources. You don't want to sell your land, but you also have to figure out how to make it pay. So you begin to look at options for development. In doing so, you discover two things. First, you discover that you can sell a part of your land to a real estate developer who will turn it into suburban housing at a pretty handsome profit. Second, you discover that your land is home to a variety of endangered species, which, under the Endangered Species Act (ESA), it is illegal to harm.

Now, nobody except you and your consultants knows that these species are on your land, and you know that if the government finds out, your development options will be severely limited (perhaps even shattered). Clearly, in this situation having the endangered species is not so much an asset as it is a very clear and present liability. If you lacked scruples – or maybe just because your financial need was great – you would be tempted to adopt what Craig Denisoff (see Chapter 8) calls the 'Three S's Approach to Species Management': Shoot, Shovel and Shut Up.

We're not saying that people around the US are busy killing endangered species as a result of the ESA. We just don't know. What we do know is that the incentive structure unwittingly set up by the ESA could push people to do just that. Fortunately, these perverse incentives are being reversed as a result of species banking. In the situation above, once species banking is in place, the landowner has a third choice: he can turn the land into a conservation bank, generate species credits, and sell those to create an income. In other words, the species that was once a liability has now become an asset.

But that is a hypothetical situation, and reality is much more complicated than can be conveyed by any theoretical story. So let us leave you with another, very real story.

This story is set in California, in the Central Valley, one of the most productive, and profitable, farming places on Earth. A strip of land that produces an unimaginable proportion of the fruits and vegetables that feed the US. A strip of land in a state whose population and economic growth has been nothing short of stunning. So much so, that the state, on its own, has now become the world's fourth largest economy, right after the UK. In other words, a state hungry for infrastructure; roads, highways, energy, all the trappings of a 21st-century economy.

Once upon a time in California there was a businessman who liked to hunt. So, together with some friends and investors, this businessman created his own small version of paradise: a hunting club on a wetland specifically designed to attract ducks and other waterfowl.

Unfortunately, the wetlands that served as a rest-stop on the flyway of our businessman's beloved ducks were also scheduled to serve as the site for a major highway bypass. In other words, it looked likely his ducks would be sacrificed to the 'greater good' of the American car. There was little that could be done: the US constitution has a provision, known as 'eminent domain' that allows the government to expropriate private property (after paying a fair amount) for the greater good. Our businessman could not refuse to sell.

Naturally, he was furious, and he probably could have sued the government and bought himself some time, not to mention perhaps a bit of money. But rather than focus on the bitterness of this particular lemon, our businessman decided to see how he could turn it into lemonade. In discussions with the local authorities, he found out that, while it was perfectly legal for the US government to strip him of his duck-hunting grounds in order to make a highway, it was not legal – thanks to the US Clean Water Act – for the government to damage a wetland without 'minimizing and mitigating' (or offsetting) that damage. In other words, the CWA states that anyone who damages a major wetland considered to be of national importance, whether it be a private landowner or the government department of transportation, needs to offset that damage by 'creating, enhancing, or restoring' a wetland 'of similar functions and values' in that same watershed.

To put it another way, the CWA, a law passed in the 1970s, has provisions that are designed to maintain a particular type of ecosystem (i.e. wetlands) because of the functions and values they provide. In other words, it has a 'biodiversity protection' provision for habitat and the associated species. This is nothing new. Many countries have laws designed to protect biodiversity. The difference, however, is that the CWA 'requires' that any damage to this particular aspect of biodiversity (wetlands ecosystems) be offset, that is, mitigated.

Which brings us back to our businessman and his ability to make lemonade out of his given lemons. When this businessman found out that in order to build the highway bypass on his wetlands, the government would have to go out and find some way of 'creating, enhancing, and restoring' wetlands of similar functions and values in that same watershed, his business instincts kicked in and he promptly went across the street and bought 315 acres of his neighbour's wetlands. He then 'enhanced and restored' this wetland complex (thus attracting ducks), went to the part of the US government that oversees the CWA (namely the US Army Corps of Engineers (USACE) and the US Environmental Protection Agency (EPA)) and got approval to sell 'wetland mitigation credits'. He then sold these mitigation credits to the government department that had taken over his land for tens of thousands of dollars an acre. Our businessman had gone from being a hunter to becoming the first 'wetland mitigation banker' west of the Mississippi.

This transformation was possible because of the CWA regulation that enabled wetland banking. As Deborah Mead explains in Chapter 2, wetland banking paved the way and helped give rise to species banking. In the mid-1990s it became clear that the US ESA could also act as a driver for species banking in the same way that the CWA made wetland banking possible. Seeing the need for species mitigation credits, our businessman and his wetland bank underwent yet another transformation. He worked with the US Fish and Wildlife Service (USFWS) to approve his bank to sell credits for vernal pool tadpole shrimp, vernal pool fairy shrimp and valley elderberry longhorn beetle. He was now a species conservation banker as well as a wetland mitigation banker.

Like many hunters, he feels a special connection to nature, to wetlands and to the animals that make his chosen sport possible. In this particular case, however, this hunters' sensibility is complemented by an innate instinct for business. For this reason, our businessman quickly realized that mitigation banking could help him marry his two loves: nature and business. And that is why, following the success of his first mitigation bank, he went on to create many more species and wetland banks, and, ultimately, established a multi-million dollar company whose sole job is to 'create, enhance, and restore' wetlands and other habitat for endangered species.

Two decades later, this businessman's company employs some 100 people and manages thousands of acres of restored wetlands and endangered species habitat. And in March of 2007, that company (known to many as Wildlands Inc.) announced that it had received a major capital infusion from Parthenon Capital, a private equity investment firm that manages more than $1.5 billion dollars. The businessman's name is Steve Morgan, and his hunter's misfortune has yielded a bit of biodiversity 'lemonade' that has turned quite sweet.

As far as Morgan is concerned, the CWA and the ESA are yielding some very unanticipated outcomes: the creation of new businesses, the sole purpose of which is the protection of biodiversity. It is wonderful alchemy; as a result of wetland mitigation banking and species banking, endangered species and ecosystems have gone from being financial liabilities to being valuable business assets.

And Steve Morgan's story is not unique; like him, there are now dozens of entrepreneurs interested in building businesses based on species and wetland mitigation. Not all will be as successful as Morgan, but the fact that there are so many out there trying (the last meeting of the US National Mitigation Banking Association brought together more than 400 participants) bodes well for the various forms of biodiversity that are eligible for banking.

No one really knows how big US wetland mitigation banking market is, but at the Ecosystem Marketplace we know that there are more than 400 wetland mitigation banks across the US and we estimate (though guess is a better word) that this market transacts more than $1 billion dollars a year. Additionally, there are now more than 70 active species banks throughout the US for red-cockaded woodpeckers, tiger salamanders, vernal pool fairy shrimp, Swainsons' hawk, California gnatcatcher and others. There are even species banks aimed at conserving flies: namely the Delhi Sands Flower-Loving Fly (going price for fly habitat: $100,000 an acre).

If all of this sounds strange, or if you are wondering how this relates to the role that business can/should play in the protection of biodiversity, here is the connection: The reality of the human condition is that the more we develop, the more impact we have on biodiversity, and the more degraded our own health and livelihood becomes. Whether it is roads and highways, cities and buildings, or even the expansion of soybean fields or palm oil fields, our impact on biodiversity is great and growing greater still. In the past, our economic system has been 'blind' to this impact. We have not taken into consideration its true cost, nor has the 'the true price' for damaging biodiversity been incorporated into the cost of houses, shopping malls or roads. With mitigation and conservation banking – and the concept of biodiversity offsets more generally (Chapter 13) – this cost is beginning to be 'internalized'. Roads that damage wetlands cost more than roads that do not, shopping malls on endangered hawk habitat cost more than shopping malls without this impact, and so on. And the more we can incorporate these true costs into our economic system, the better off biodiversity will be.

If there is one thing we've learned about businesses, it is that they are very good at managing things that show up in their economic equations as costs. So if the cost of biodiversity can be made more obvious, perhaps it will be better managed. The converse is also true: if businesses see a financial opportunity in conserving biodiversity, or if money can be made protecting species and habitat, then chances are better still that it will be protected.

We have three objectives for this book. The first is to use species credit banking (also known as conservation banking) in the US as a model to demonstrate how an active biodiversity market can and might function. The 'Overview' part (Chapters 1, 2, 3 and 4) provides an introduction to this model. The second objective is to provide practical guidance to practitioners and those interested in establishing a bank in the US, or a banking system outside the country. The 'Establishing a Conservation Bank' part covers ecological (Chapter 5), legal (Chapter 6), regulatory (Chapter 7), business (Chapter 8) and financial (Chapter 9) considerations. Finally, we hope to give readers a peak at where we think conservation banking may be heading in the near, and not-so-near, future. The final parts on the 'State of the Art' (Chapters 10, 11 and 12), 'Going Global' (Chapters 13 and 14) and 'Conclusion' (Chapter 15) address this objective.

We hope that this book will serve to help the 'Steve Morgans' of the future to navigate the complexities of the US market and, beyond that, to help practitioners around the world see how systems such as this might (or might not) work in places as far afield as Australia and Ghana. Finally, we hope you enjoy the book, and that is spurs you to work with us (and the other authors in this book) to help biodiversity markets and biodiversity offsets become a truly effective tool for biodiversity conservation.

2

History and Theory: The Origin and Evolution of Conservation Banking

Deborah L. Mead

Environmental legislation creates new markets

Anyone familiar with conservation banking will tell you its roots are in wetland mitigation banking. Wetland mitigation banking is a regulated system that allows those impacting wetlands to fund the creation, restoration or enhancement of wetland habitat at one location, a mitigation bank, as a means of offsetting the loss of the same type of wetland at multiple other locations (i.e. the impact sites). The offset habitat at the wetland mitigation bank is represented by wetland credits, which are purchased by developers or others to compensate for their projects' impacts to this same type of wetland habitat elsewhere. Conservation banking is a similar concept where adverse impacts to endangered, threatened or other protected species are offset at a conservation bank where the credits represent individuals or habitat.

In 1992 the Coles Levee Ecosystem Preserve was established in Kern County, California, as mitigation for Atlantic Richfield Company's oil and gas development activities. Although not named a 'conservation bank', this 6059-acre preserve functioned as a conservation bank for its owner and later provided credits for sale to others with impacts to the San Joaquin kit fox (*Vulpes macrotis mutica*), Tipton kangaroo rat (*Dipodomys nitratoides nitratoides*), blunt-nosed leopard lizard (*Gambelia sila*) and other upland species in the San Joaquin Valley. This conservation bank is now owned by Aera Energy, LLC, which has retained the remaining credits for internal use. It is one of the largest conservation banks in California.

Carlsbad Highland Conservation Bank is generally recognized as the first official conservation bank, because it was established in coordination with the announcement of California's 'Official policy on conservation banks' (Wheeler and Strock, 1995). This parcel of coastal California gnatcatcher (*Polioptila californica californica*) habitat located in southern California was approved as a conservation bank by both the United States Fish and Wildlife Service (USFWS) and the California Department of Fish and Game (CDFG) in April 1995. The property was owned by the Bank of America, which obtained it through a foreclosure (Environmental Defense

Fund, 1999). Its appraisal value was low, primarily because the property consisted of coastal sage scrub, habitat of the gnatcatcher, which was listed as a threatened species by USFWS in March 1993. The Bank of America realized the property had limited development potential and instead looked for opportunities to profit from its ecological value (Environmental Defense Fund, 1999). Eighty-three acres were sold to California's department of transportation for use as mitigation for impacts to the gnatcatcher and the remaining 180 acres were developed into the Carlsbad Highlands Conservation Bank. This bank was the first of many associated with a large, regional conservation plan in southern California, the San Diego Multiple Species Conservation Plan (MSCP).

The mitigation banking concept began in the 1970s, with the enactment of the Federal Water Pollution Control Act of 1972 (33 U.S.C. 1251 et seq.), commonly referred to as the Clean Water Act (CWA). The CWA provides for protection of wetlands and waters of the US and is administered by the United States Army Corps of Engineers (USACE) with oversight by the Environmental Protection Agency (EPA). The CWA section 404 permit program and the wetland conservation provisions of the Food Security Act, which is administered by the Natural Resource Conservation Service (NRCS), require mitigation for projects that will impact wetlands. Compensatory mitigation for these projects is traditionally conducted project-by-project, resulting in numerous, small mitigation sites on or near the properties under development. A similar project-by-project approach is primarily in use for offsetting impacts to endangered and threatened species. This approach has not proven very successful, as described later in this chapter.

To address this problem, in part, USACE began approving wetland mitigation banks in the 1980s, creating a market for wetland mitigation credits. By the mid-1990s more than 50 wetland mitigation banks (Marsh et al, 1996) and a few conservation banks had been established. In 1995 'Federal guidance for the establishment, use and operation of mitigation banks' was issued by USACE, EPA, NRCS, USFWS, and the National Oceanic and Atmospheric Administration's National Marine Fisheries Service (NOAA Fisheries) (*Federal Register* 60: 58605–58614). This interagency guidance, although directed mainly at wetland mitigation banking, provided conceptual and procedural guidance useful for conservation banking.

The Endangered Species Act of 1973 (16 U.S.C. 1531 et seq) (ESA) also opened up new markets. Mitigation banks for endangered or threatened (i.e. listed) species or other at-risk species are often referred to as conservation banks. Legal requirements, most notably the ESA, concerning protected species' habitat underpin most conservation banking activity. The ESA aims to: (1) conserve the ecosystems upon which listed species depend, and (2) provide a programme for the conservation of these species. The term 'conserve' is defined in the ESA as the use of all methods and procedures which are necessary to bring any endangered or threatened species to the point at which the measures provided pursuant to the ESA are no longer necessary. While no explicit mention of conservation banking is found in this statute or its implementing regulations, the law does require that unavoidable impacts to listed species be minimized or mitigated. Such measures do not preclude the use of off-site mitigation, like that which occurs through conservation banking. Specifically, it is the authorized incidental take of federally listed species under sections 7 or 10 of the ESA that provides the demand for most species conservation credits. For an explanation of incidental take authorization, see Box 2.2. The ESA is administered by USFWS and NOAA Fisheries.

California is where conservation banking began. It was formally launched on 7 April 1995, when the California Resources Agency and the California Environmental Protection Agency jointly issued their 'Official policy on conservation banks' (Wheeler and Strock, 1995). This policy set standards for bank site selection, use of legally enforceable implementing agreements, use of conservation easements and provisions for long-term management and monitoring. In January 1996 CDFG and USFWS issued the 'Supplemental policy regarding conservation banks within the natural community conservation planning area of southern California'. This supplemental policy provided further guidance regarding bank size, function, credits and service areas. Furthermore, and perhaps most importantly, it expressed support for conservation banking and a free market approach to mitigation within this large planning area. More specifically, the guidance directed that, 'the number of conservation banks that are established will be regulated by the free market and willingness of landowners to participate, not by the wildlife agencies'.

In addition to the ESA, the California Endangered Species Act (CESA) and the California Environmental Quality Act (CEQA) provide a legal nexus for species and habitat conservation banking in California. The CESA provides protection for animal and plant species listed by the State of California and functions somewhat similarly to the ESA. The CEQA is a protection act which requires that significant effects to natural resources, including species, be mitigated to a non-significant level. The CEQA acts as a safety net for species and habitats that are not protected under the CWA, ESA, CESA or other environmental protection statutes. Together these laws and their implementing regulations drive a mitigation banking market for a variety of natural resources in California. See Box 2.1 for a timeline of milestone events in conservation banking.

Box 2.1 *Mitigation banking milestones*

The events listed below had a significant part in laying the foundation for, establishing and developing conservation banking as a natural resource conservation tool in the US.

April 1970 The first Earth Day was held on 22 April. This event was the culmination of a decade of environmental activism and planning and spurred the creation of much environmental legislation in the US.

January 1971 The California Environmental Quality Act (California Public Resources Code section 21000 et seq) was signed into law in 1970 and took effect 1 January 1971, exactly one year after the National Environmental Policy Act (NEPA) was enacted.

October 1972 The Federal Water Pollution Control Act (33 U.S.C. 1251 et seq) was signed into law.

December 1973 The Endangered Species Act (16 U.S.C. 1531 et seq) was signed into law.

October 1982 Section 10 of the ESA was amended to include the use of Habitat Conservation Plans (HCPs), thus allowing mitigated, incidental take of federally listed species by private landowners.

June 1983 USFWS issued interim guidance on mitigation banking through a memorandum to its regional directors that recognized the potential for mitigation banking.

January 1984 The first wetland mitigation bank, Tenneco LaTerre Mitigation Bank, was established in Louisiana through a Memorandum of Agreement with USACE (McElfish and Nicholas, 1996).

January 1985 The current California Endangered Species Act of 1984 (Fish and Game Code section 1050 et seq) took effect. It superseded the state's original 1970 Act and now closely resembles the federal ESA.

Early 1990s USACE and EPA issued Memorandums of Agreement regarding mitigation banking. A number of regional guidance documents from various federal agency offices clarified agency expectations regarding mitigation banking and officially sanctioned their use.

April 1991 The California Natural Community Conservation Plan (NCCP) Act (Fish and Game Code, section 2800 et seq) was passed.

October 1992 Coles Levee Ecosystem Preserve was established in Kern County, California, by the Atlantic Richfield Company primarily as a conservation bank for its own use to offset oil and gas development activities.

1994–1999 Several reports were published describing the general failure of the project-by-project mitigation approach (DeWeese, 1994; Marsh et al, 1996; Redmond et al, 1996; Environmental Defense Fund, 1999), prompting agencies to take a mitigation banking approach.

April 1995 The State of California announced its 'Official policy on conservation banks' (Wheeler and Strock, 1995).

April 1995 The first official species conservation bank, Carlsbad Highlands Conservation Bank, located in southern California, was approved by USFWS and CDFG for use in association with the San Diego Multiple Species Conservation Plan, the first HCP–NCCP.

November 1995 'Federal guidance for the establishment, use and operation of mitigation banks' was issued by USACE, EPA, NRCS, USFWS and NOAA Fisheries (*Federal Register* 60: 58605-58614).

January 1996 CDFG and USFWS issued the 'Supplemental policy regarding conservation banks within the NCCP area of southern California'.

February 1996 USFWS issued the first programmatic formal ESA section 7 consultation specifically promoting the use of species conservation banks.

April 1997 The Hawaii Legislature passed a bill amending its state endangered species act (Haw. Rev. Stat. section 195D-21(b)) to allow for habitat banking through HCPs.

July 1999 USFWS issued 'Method for determining the number of available vernal pool preservation credits in ESA conservation banks in the California Central Valley', providing guidance on conservation banking.

November 2000 'Federal guidance on the use of in-lieu-fee arrangements for compensatory mitigation under section 404 of the Clean Water Act and section 10 of the Rivers and Harbors Act' was issued (*Federal Register* 65: 66914–66917).

May 2003 'Guidance for the establishment, use, and operation of conservation banks' was issued by USFWS.

Box 2.2 *Endangered Species Act incidental take authorization*

The Endangered Species Act is administered by NOAA Fisheries where marine and anadromous species are concerned (with some exceptions), and by USFWS where all other species are concerned. Section 9 of the ESA prohibits take of endangered fish and wildlife species. 'Take' means to harass, harm, pursue, hunt, shoot, wound, kill, trap, capture or collect one or more individuals of a listed species, or to attempt to engage in any such conduct. 'Incidental take' is defined in the ESA as take of a listed species that is incidental to, and not the purpose of, carrying out an otherwise lawful activity. Activities such as building a home, constructing a bridge or converting range or crop land to more intensive agricultural uses that are likely to adversely affect listed species, require incidental take authorization from USFWS, NOAA Fisheries or both agencies, depending on the species. Incidental take of listed species can only be authorized through an incidental take statement (ITS) issued under section 7 or an incidental take permit (ITP) issued under section 10 of the ESA.

Federal agency actions that may affect listed species consult with USFWS/NOAA Fisheries on the proposed action under section 7(a)(2) of the ESA. If the proposed action 'is likely to adversely affect' one or more listed species, formal consultation is required. Formal consultation culminates in USFWS/NOAA Fisheries issuing: (1) a biological opinion with a finding that the proposed project is or is not likely to jeopardize the continued survival or recovery of one or more listed species, and a finding that the proposed project is or is not likely to destroy or adversely modify designated critical habitat; and (2) in the case of a non-jeopardy biological opinion that will not destroy or modify critical habitat, an ITS exempting a specified maximum amount of incidental take of one or more listed species.

Projects proposed on private lands that (1) require a permit from a federal agency, for example a building project that requires a USACE permit to fill wetlands, or (2) intend to use federal funds, for example a state department of transportation project that will use Federal Highway Administration funds, have a federal nexus (a connection to a federal agency, which is considered the action agency), and must acquire incidental take authorization through that federal nexus under section 7 of the ESA. It is the action agency that receives incidental take authority from USFWS/NOAA Fisheries and extends that authority to the private party through the action (e.g. USACE issuing a permit to fill wetlands protected under the CWA). Many private project proponents receive incidental take authorization through this process.

Private project proponents without a federal nexus cannot use section 7 and instead develop a HCP and apply for an ITP, which is issued under section 10(a)(1)(B) of the ESA. A HCP can include both listed and non-listed species, but must include at least one listed species. Many project proponents proposing long-term projects request coverage for non-listed species that they think are at risk of becoming listed prior to the expiration of the ITP for which they are applying. The term of an ITP is a function of the proposed action and must be of sufficient duration to cover all incidental take of all covered species through completion of the project and the successful implementation of the mitigation measures. Large multi-species plans, such as countywide HCPs, often request incidental take coverage for many species over many years. For example, the ITP for the San Diego MSCP covers 85 species for 50 years. The HCP process also involves a set of findings, including an internal consultation under section 7, and culminates in (1) approval or non-approval of the HCP, and (2) an ITP for approved HCPs that authorizes a specified maximum amount of incidental take of one or more listed species.

In both cases project proponents often propose compensatory habitat to minimize or mitigate unavoidable impacts, primarily habitat loss, to species. The purchase of credits from a conservation bank to offset these species' habitat losses is almost always the first choice of project proponents when credits are available. Mitigating through the purchase of conservation bank credits transfers the responsibility for success of the mitigation from the project proponent to the conservation bank owner.

Conservation banking: Definition and how it works

Before discussing the details of conservation banking, it is important to understand the underlying concept of mitigation. The term mitigation is defined differently in various federal and state laws, regulations and policies. For example, the definition of mitigation in the National Environmental Policy Act (NEPA) is different than that used by USACE in implementing the CWA or USFWS in implementing the ESA. In the broad sense mitigation is defined as everything from avoiding the impact through to compensating for the impact by replacing or providing substitute resources (i.e. compensatory mitigation). This is the definition used in NEPA. In the narrow sense, mitigation is defined as providing compensatory, in-kind and geographically proximate habitat which is occupied by the same type of natural resource impacted by the action. In practice, this is generally the definition used by USFWS for listed species, although the geographically proximate requirement is relaxed somewhat to allow for off-site mitigation that is environmentally favourable. A point often brought up regarding statutory language in the ESA is that section 7 of the ESA requires minimization, rather than mitigation, of the level of take of listed species. Providing compensatory habitat through the purchase of credits at a conservation bank is a method of minimizing the level of take when habitat loss is unavoidable.

Under section 404 of the CWA and section 10 of the Rivers and Harbors Act of 1899 (33 U.S.C. 403 et seq), a process known as sequencing requires that project applicants must undertake all appropriate and practicable steps to first avoid and then minimize adverse impacts to aquatic resources, prior to mitigating any loss of such resources at a wetland mitigation bank or other compensatory mitigation site. The ESA generally uses this same hierarchical approach. The purpose of conservation banking is not to encourage or facilitate development of species' habitat where it can be avoided, but to provide compensation for unavoidable impacts and avoid ecologically ineffectual on-site compensation.

Conventional mitigation for impacts to natural resources is generally accomplished on-site and in-kind. On-site means compensatory mitigation located on a portion of the project area or an adjacent parcel. In-kind means preserving, enhancing, restoring or creating the same type of habitat that was impacted by the project. Out-of-kind mitigation, compensating impacts with a different type of habitat or other natural resource from that which was impacted, is highly discouraged and rarely used. The natural resources at issue may be wetlands, federally listed species, state listed species, sensitive habitat types or other resources that are protected by law. Such on-site, compensatory mitigation conducted in the past on a project-by-project basis collectively consists of many small, disjunct mitigation sites. Historically, these mitigation sites have little or no long-term management requirement on the part of the landowners, cannot be adequately defended from surrounding incompatible land uses, are inadequately monitored for compliance and effectiveness of the mitigation and rarely serve their intended long-term purposes.

Conservation banking is essentially in-kind, off-site mitigation in which multiple projects with like impacts are mitigated at the same location, the conservation bank. Off-site means compensatory habitat preserved, enhanced, restored or created at a location other than the project site. The ability to focus conservation efforts on fewer, larger, strategically located mitigation sites which are ideally selected for their high environmental value to the affected natural resources at issue, and which have the advantage of economy of scale, is thought to greatly increase the chances that the mitigation will be successful. Conservation banking avoids the problems associated with conventional mitigation, in which small mitigation parcels are scattered across the landscape, making it difficult to effectively monitor, manage or defend these sites and ensure that they provide their intended environmental functions over time.

Conservation banking is a term first introduced in California's official conservation banking policy, distinguishing non-wetland mitigation banks established for natural resource values other than wetlands from wetland mitigation banks (Wheeler and Strock, 1995). As defined by CDFG, a conservation bank is privately or publicly owned land managed for its natural resource values. In exchange for permanently protecting the land, the bank owner is allowed to sell habitat credits to parties who need to satisfy legal requirements for compensating environmental impacts of development projects.

USFWS adopted the term conservation bank to distinguish banks for federally listed species from wetland mitigation banks. USFWS defines a conservation bank as a parcel of land containing natural resource values that are conserved and managed in perpetuity, through a conservation easement held by an entity responsible for enforcing the terms of the easement, for specified listed species and used to offset impacts occurring elsewhere to the same resource values on non-bank lands (USFWS, 2003).

Both the CDFG and USFWS definitions are conceptually and functionally similar. There are three important elements within both definitions. The first is protection and management of the bank lands in perpetuity. Without the long-term assurances a perpetual conservation easement offers, mitigation cannot truly compensate for permanent impacts to listed and other at-risk species and their habitats. Property use restrictions without an easement, such as deed restrictions, are not as legally binding as a conservation easement and tend to be lost over time as property titles change.

The second important element is management in perpetuity and the underlying implication that adequate funding will be provided to accomplish this task. It has long been recognized in the field of conservation biology that few, if any, preserved parcels can retain their initially intended function without managed intervention (e.g. prescribed burning, prescribed grazing, alien invasive species control, maintenance of a succession state of vegetation). Basic costs usually include: funds for repair and replacement of fences, signs and equipment necessary to manage the site; fees for biological monitoring and reporting; taxes and insurance. Furthermore, the use of contingency plans and funding to address unexpected or infrequent threats (e.g. a new alien species invasion, flooding or fire) has proven necessary.

The third important element inherent in these definitions is that conservation banking is a free market enterprise that allows for the sale, purchase or trade of habitat or species, represented by a currency referred to as credits. Various legally protected natural resources are represented by credits which are awarded by the resource agencies to bank owners. Bank owners can sell these credits to others that are required to mitigate project impacts on the same type of natural resources elsewhere. As in any market, there must be both supply and demand for such a system to work. Conservation banking credits are the economic reward supplied by the resource agencies to the landowner who agrees to preserve, protect and manage habitat in perpetuity. The demand for these credits is driven by development and enforcement of environmental laws or policies that require compensatory mitigation for adverse impacts to these same types of natural resources elsewhere. Market prices are based on how much buyers will pay. Since conservation banking is voluntary, there is a built-in cap on prices – buyers are unlikely to pay more than what it would cost them to supply the required mitigation on their own. See Appendix IV for a template of a conservation banking transaction.

There are two major ecological advantages of conservation banking. First, because conservation banks are established ahead of project impacts to species and their habitats, there is a reduction in temporal loss of habitat (i.e. temporary disturbance or loss of habitat that is reclaimed in the future). Temporal loss of habitat is almost always associated with project-by-project mitigation, because mitigation sites are usually established coincidentally with project impacts. In addition, credits for restored or created habitats in conservation banks are generally released on a schedule commensurate with the success of the restored or constructed habitat. Specific performance criteria are agreed to upfront by the agencies and the bank owner. This greatly reduces the risk associated with failed mitigation because the agreed-upon performance criteria must be met before credits are released. This process generally leads to more carefully planned and executed mitigation projects because failure by the bank owner to achieve the performance criteria usually results in reduced profits.

The second major ecological advantage of conservation banking over project-by-project mitigation is that it is usually part of a larger conservation strategy. The supposition is that agencies and communities that plan mitigation efforts in advance of the impacts have a better chance of setting aside more environmentally important parcels and can better incorporate mitigation into a preserve system that will allow for sustainable populations of listed and other at-risk species. Recall that one of the goals of the ESA is to conserve the species and the ecosystems on which they depend. Engaging in regional conservation planning ahead of development impacts and selecting conservation bank sites and other conservation areas before habitat loss and fragmentation occur can increase the chances of successfully accomplishing this goal. HCPs, which are described in more detail below, use this approach. For more on the advantages of conservation banking see Chapter 3.

Proceeding with caution: Lessons learned from past mitigation efforts

Conservation banking was slow to start because market conditions did not exist during the first two decades the ESA was in place. Very little compensatory mitigation was proposed or required to offset impacts to listed species during this time. As both the amount and rate of urban, industrial, rural and agricultural development increased and the associated adverse impacts to listed species increased, it became obvious that a strategy that did not include compensatory mitigation for destroyed or otherwise permanently impacted habitat would preclude recovery of most listed species. In addition, relatively few species, approximately 200, were listed at the time the ESA was enacted. As of this publication, there are more than 1300 species listed as threatened or endangered, and about 280 proposed or candidate species for listing (USFWS, 2007). Petitions to list additional species are submitted to USFWS and NOAA Fisheries each year.

Uncertainty on the part of resource agencies, developers and prospective entrepreneurial bankers also contributed to conservation banking's slow start. Uncertainty on the part of the agencies was mainly due to the lack of success demonstrated by early wetland restoration and creation projects (DeWeese, 1994; Marsh et al, 1996; Redmond et al, 1996; Environmental Defense Fund, 1999). Unfortunately most of the early mitigation sites failed or provided much less than the expected ecological functions and values needed to offset the impacts for which they were intended to compensate. A study of wetland creation projects in California (DeWeese, 1994) found that while the goal of no-net-loss of wetlands was generally met in three to five years, replacement of in-kind function and values of the various types of created wetlands was not nearly as successful during this same period and, on average, none achieved high value. In addition to lack of compliance (i.e. failure to implement the creation or restoration project as planned) and lack of success (i.e. failure to meet the stated biological goals and objectives), a critical concern was found to be the temporal loss associated with compensatory mitigation of this type. The time required for some restored or created habitat types to reach a stage at which there is even the potential to provide the intended ecological functions was generally greater than five years and, in the case of riparian habitats, greater than 25 years. In addition, creation of wetland habitat was found to be much less successful than restoration of wetland habitat (Redmond et al, 1996).

The project-by-project mitigation approach for the threatened valley elderberry longhorn beetle, *Desmocerus californicus dimorphus*, a species completely dependent on elderberry, *Sambucus* spp., was found to be generally unsuccessful for some of the same reasons cited above (G. R. Huxel and S. K. Collinge, unpublished manuscript). Non-compliance, unsuccessful habitat restoration and lack of colonization of successfully restored habitat by the beetle were the primary reasons for lack of success. For this species the use of conservation banks, which are strategically sited within short distances of existing, occupied habitat, was found to be environmentally preferable (G. R. Huxel and S. K. Collinge, unpublished manuscript; Huxel and Hastings, 1999).

Compounding the uncertainty surrounding the largely unsuccessful project-by-project mitigation approach for both wetland and upland habitat types, many of the early wetland mitigation banking efforts also failed, usually for the same reasons the individual project-by-project mitigation sites failed. Mitigation bank failure is of even greater concern due to the fact that multiple projects are mitigated at a single bank site and there is potentially more to lose should failure occur.

The reasons for the lack of success are varied, but generally can be attributed to one or more of the following circumstances:

- Failure to implement the mitigation.
- Mitigation sites too small to be ecologically sustainable.
- Absent or insufficient management of the mitigation site.
- Absent or insufficient funding for long term management of the mitigation site.
- Faulty engineering and/or construction of habitat in the case of restored or created habitats.
- No contingency plan or funding for unexpected or infrequent events (e.g. natural catastrophes or new alien species invasions).
- Lack of support for the mitigation site by the community which often results in unchecked vandalism, unauthorized off-road-vehicle use, and dumping on the site.

Unfortunately, under many wetland mitigation programmes, once the mitigation site was given a passing grade and signed off by the regulatory agencies, the site ceased to receive further attention and quickly deteriorated. In general, long-term monitoring and management needed to maintain a functioning mitigation site after success criteria were met was not required by the agencies.

This situation resulted in agency reluctance to provide assurances to prospective bank owners, for example, guaranteeing the release of a fixed number of credits within a specified period, creating uncertainty on the part of both prospective bank owners and potential credit purchasers. The lack of agency assurances was perceived by many prospective bank owners as a significant business risk and constrained them from taking risks associated with establishing a mitigation bank for which they may not receive enough credits or enough agency support to market the credits successfully. In the early 1990s a number of guidance documents from various federal agency offices better clarified agency expectations regarding mitigation banking and officially sanctioned the use of mitigation banks, alleviating some of the concerns raised by prospective mitigation bank owners.

The first conservation bank proposals were slow to be approved. Should a conservation bank fail, recovery for the species would likely take a step backward and the risk of jeopardizing a species' recovery was a real concern. Agency biologists worked closely with prospective bank owners and their biological consultants and attorneys, sometimes for years, striving to overcome the shortcomings evident at project-by-project mitigation sites and avoid problems experienced by some previous wetland mitigation banking attempts. These efforts paid off and most conservation banks appear to be meeting their intended goals. Although detailed biological studies of the effectiveness of conservation banking have yet to be conducted, annual reports from conservation managers, easement holders and agency biologists appear to indicate that generally conservation banking is an ecologically successful method for offsetting impacts to many species.

Many lessons have been learned from previous mitigation successes and failures, both biological and procedural. For the most part, these lessons have been incorporated into the wetland mitigation and conservation banking policies and guidelines issued by various federal and state agencies. Some of these lessons are reflected in USFWS requirements for conservation bank approval and are described later in this chapter.

The HCP–NCCP–conservation banking connection

In 1982, section 10 of the ESA was amended to include the use of HCPs by private landowners, corporations, tribal governments, state and local governments and other non-federal landowners as a method to obtain authorization for incidental take of listed species. Previous to this change in the ESA, only federal agencies or private landowners with a federal nexus could get authorization for incidental take of listed species. See Box 2.2 for an explanation of incidental take of listed species under section 7 and 10 of the ESA. A HCP describes the anticipated, unavoidable impacts on species by the proposed project, including the type and amount of incidental take of each species to be covered under the plan, and how those impacts will be minimized and mitigated. Project proponents are encouraged to address both listed and unlisted species in HCPs. All non-federal landowners without a federal nexus who want an incidental take permit must prepare a HCP and apply for an incidental take permit.

Species conservation strategies or frameworks are generally developed shortly after species are federally listed. These frameworks are often used in the development of a conservation banking programme. Large, regional HCPs further develop these conservation frameworks as part of the HCP process, identifying the needs of listed and non-listed species of fish, wildlife and plants in the plan area that will be covered by the HCP, threats to these resources, and measures that can be implemented to conserve these resources. Identified conservation measures often include the use of preserves and conservation banking. This approach allows species conservation to be proactive and for mitigation to occur ahead of the impacts. Such up-front planning affords communities the opportunity to reduce habitat fragmentation, set aside more environmentally important parcels, and design a preserve system with wildlife corridors and other features that allow for sustainable populations of listed and non-listed species, greater conservation of species' genetic variation and increased habitat diversity.

A NCCP is the California version of a large HCP. Because NCCPs are used only for large planning efforts such as countywide plans, and California has more than 300 federally listed species and additional state listed species, a NCCP is almost always done jointly with a HCP. Due to the large scale on which conservation planning occurs for HCP–NCCPs, conservation banking is often included in these plans. An important feature of HCPs and NCCPs is that plan proponents are encouraged, although not required, to design conservation strategies that will provide net conservation benefits for species.

The NCCP program was the precursor to California's conservation banking programme. The San Diego MSCP was the first NCCP and the first HCP–NCCP, and it is this large, multi-species plan that launched conservation banking in southern California. The San Diego MSCP consists of an umbrella HCP–NCCP, under which there will be several sub-area HCP–NCCPs, at least six of which have been completed (City of San Diego, 2007). The San Diego MSCP has made extensive use of conservation banking, especially for the coastal California gnatcatcher and other co-occurring species within the coastal sage scrub community. Some of these banks have been awarded multi-species credits for coastal sage scrub habitat (see Chapter 5 and Chapter 7 for further discussion on credits). Approximately 15 conservation banks are associated with the San Diego MSCP.

In recent years large-scale, multi-species HCPs are being used to direct conservation efforts in fast developing areas. Many of these HCPs are using, or plan to use, conservation banking or a conservation banking approach as part of their conservation strategy, including most of the countywide HCPs recently approved or pending approval in the western US. This level of planning strives to achieve upfront conservation for development impacts, net conservation benefits for species and streamlined regulatory processes associated with the ESA, CWA and other federal, state and local environmental laws.

ESA vernal pool crustaceans listings lead to conservation banks throughout California

In northern and central California, conservation banking at the federal level was launched by the listing of several vernal pool species, most notably the vernal pool fairy shrimp (*Branchinecta lynchi*) and the vernal pool tadpole shrimp (*Lepidurus packardi*) (*Federal Register* 59: 48136–48153). Vernal pools are ephemeral wetlands that in California fill with winter rains and then dry up through the spring and early summer. They are essentially tiny islands teaming with native species that dot a grassland landscape composed mostly of non-native plant species. Typical inhabitants of vernal pools are endemic, and often rare, plants and invertebrates adapted to this harsh, seasonally wet environment that is flooded for a short time each year and dry the rest of the year. While some protection is afforded vernal pools through the CWA, it was not sufficient to protect many of the species that depend on these wetlands and their associated uplands.

In contrast to the southern California conservation banks which have been established primarily in association with the San Diego MSCP through section 10 of the ESA, most conservation banks in central and northern California have been established in association with section 7 of the ESA. Due to the dependence on wetland or stream habitats by many of the species listed in these areas, impacts to their habitats often trigger the need for a permit from USACE. See Box 2.2 for an explanation of incidental take authorization under section 7 of the ESA.

To better conserve the listed vernal pool species and handle the greatly increased ESA section 7 consultation workload that resulted from these species' listings, USFWS developed a conservation strategy in coordination with USACE and other public and private stakeholders. A four-pronged approach was used that included:

1 issuance of a programmatic consultation to which projects could be appended, reducing the consultation period from 135 to 30 days for small impact projects;

2 identification of target areas for preservation within different California vernal pool regions;
3 incentives for landowners with ecologically valuable habitat within these target areas to establish conservation banks on their lands;
4 incentives for project proponents to use these conservation banks.

The 'Programmatic formal Endangered Species Act consultation on issuance of [CWA section] 404 permits for projects with relatively small effects on listed vernal pool crustaceans within the jurisdiction of the Service's Sacramento field office', issued on 19 February 1996 (Service file 1-1-96-F-1), greatly facilitated the ESA consultation process for USFWS, USACE and many project proponents seeking permits from USACE within the Central Valley of California. This programmatic consultation established mitigation ratios for habitat preservation and habitat creation (i.e. enhancement, restoration or new construction of vernal pool habitat) to offset impacts to vernal pools by small-impact projects. Still in effect, this programmatic consultation offers lower mitigation ratios to project proponents who choose to mitigate through the purchase of credits in USFWS-approved conservation banks rather than attempting to accomplish compensatory mitigation on their own. Participation in the programmatic consultation is voluntary and project proponents can instead choose to submit an alternative project proposal and have an individual consultation. Project proponents who qualify to participate in the programmatic consultation and mitigate at a conservation bank choose to do so with very few exceptions because it is usually the most expedient and cost-effective process for getting incidental take authorization for the listed vernal pool crustacean species.

Encouraging landowners to establish conservation banks to service the demand for mitigation credits generated by the programmatic consultation was achieved, in part, through: (1) issuance of the 'Method for determining the number of available vernal pool preservation credits in ESA conservation banks in the California Central Valley' on 26 July 1999 (an interim version was in use by late 1996); and (2) establishment of an associated in-lieu fee programme. The credit determination method was designed to target preservation of the most environmentally desirable vernal pool areas by awarding landowners with more ecologically valuable parcels a greater number of credits. It also provided basic guidance on bank requirements, crediting and service areas. Scoring criteria to determine the number of credits to be awarded for the proposed bank lands included:

- preserve size;
- vernal pool type, in which rare types scored higher;
- number of state and federally listed, proposed and candidate species occurring on the site, in which multi-species sites scored higher;
- number of other rare or at-risk species occurring on the site from a list compiled jointly by the agencies;
- condition of the site;
- defensibility of the site.

The difference between a low scoring and high scoring site could be more than twice the number of credits for the same amount of vernal pool wetlands on the same size parcel. Twice the credits can equate to twice the income for the conservation bank owner, an effective incentive for encouraging landowners with more ecologically valuable lands to become conservation bank owners. Once conservation banks had been established in most of the targeted areas, a simpler credit method in which one wetted acre of vernal pools equals one credit was adopted.

The establishment of two in-lieu fee funds, one for vernal pool preservation banks, which was held by The Nature Conservancy, and one for vernal pool creation banks, which was held by the Center for Natural Lands Management, served two purposes. First it enabled the programmatic consultation to function prior to the establishment of conservation banks, thus preventing permitting delays. This was politically important because the listing of the vernal pool crustaceans was highly controversial. The other important purpose for establishing these in-lieu fee accounts was the incentive a fund earmarked for new conservation banks provided prospective bank owners. The ability to provide a new bank owner with a guaranteed credit purchase from the in-lieu fee funds upon bank approval was often enough to offset the up-front costs incurred through establishment of the bank, making the financial venture feasible.

The use of in-lieu fee funds should be carefully considered. While is it possible to use such funds in conjunction with section 7 of the ESA, it is not possible to use in-lieu fees in place of mitigation for HCPs under section 10 of the ESA, even for a short period. In-lieu fee funds must not compete with existing conservation banks or other available, approved mitigation sites. The value of in-lieu fee funds is facilitating recruitment of new mitigation banks when a market demand for such banks exists. Also, in-lieu fees for these funds should be set to reflect the existing market and should be updated as often as needed. Legal issues or other problems may arise if in-lieu fee funds set, change or otherwise interfere with the free market. Inter-agency guidance, 'Federal guidance on the use of in-lieu-fee arrangements for compensatory mitigation under section 404 of the Clean Water Act and section 10 of the Rivers and Harbors Act', was issued in November 2000 (*Federal Register* 65: 66914–66917).

The approach described above for the listed vernal pool crustaceans proved to be very successful. The first conservation bank to provide vernal pool preservation credits, Orchard Creek Conservation Bank, opened shortly after the programmatic consultation was issued. In-lieu fees collected from this vernal pool region were directed to this new bank immediately upon its approval. More than 20 conservation or joint conservation-wetland mitigation banks for vernal pool species have been approved as part of this programme.

The success of the vernal pool species mitigation banks and streamlining of the ESA regulatory process that this approach demonstrated, led to similar approaches for other federally listed species within the jurisdiction of the Sacramento Fish and Wildlife Office including: valley elderberry longhorn beetle; California red-legged frog (*Rana aurora draytonii*), giant garter snake (*Thamnophis gigas*), San Joaquin kit fox (*Vulpes macrotis mutica*) and Alameda whipsnake (*Masticophis lateralis euryxanthus*). Conservation banks for all of these species have been established.

Federal guidance, at last!

On 2 May 2003, USFWS issued 'Guidance for the establishment, use, and operation of conservation banks' (USFWS, 2003). Approximately 45 conservation banks had been approved by USFWS prior to the issuance of this national guidance. While many of these banks have legal and financial assurances as stipulated in the guidance, some do not. Not all of these bank sites were secured with perpetual conservation easements and a few do not have sufficient long-term funding. Some were established under a Memorandum of Agreement (MOA), Memorandum of Understanding (MOU) or other agreement which does not meet the current standard, which is a Conservation Bank Agreement (CBA), Bank Enabling Instrument (BEI) or equivalent document. Despite these shortcomings, most of these conservation banks are meeting their intended purposes. In a few cases additional management, monitoring or funding, which was not part of the original agreement, has been provided either by the bank owner or other stakeholders to remediate deficiencies. In some cases, properties referred to as mitigation or conservation banks have never been formalized with a contract or conservation easement and may be vulnerable to dissolution, or may never have been intended as permanent conservation for the species.

The USFWS conservation banking guidance is similar to the inter-agency wetland mitigation guidance regarding processes, but different in its goals. The goal of wetland mitigation banking is wetland restoration, creation and enhancement to compensate unavoidable wetland losses, while the goal of conservation banking is to conserve species. Mitigation through the preservation of existing wetlands is rarely undertaken because is does not accomplish the national goal of no-net-loss of wetlands. Alternatively, mitigation aimed at conserving species often consists of preservation of existing habitat occupied by listed species and is often a preferred mitigation measure. Mitigation for listed species may take many forms depending on: (1) the needs of the species; (2) the current and projected threats to the species; (3) species recovery goals; and (4) the strategies developed to sustain populations of these species in the long term. Restoration and enhancement of habitat, and even creation of habitat in some circumstances, can be appropriate mitigation for conserving species, and all of these measures can be achieved through conservation banking.

Experience gained from establishing the early conservation banks, 'California's official policy on conservation banks' and input from individuals involved in conservation banking influenced the content of the USFWS conservation banking guidance. Summarized below are some of the basic requirements for establishing conservation banks identified in the USFWS guidance. While this guidance is specific to the US, it is generally useful for others involved in the establishment and operation of conservation banks elsewhere.

- It is advantageous to start with a conservation strategy or other framework that aids in communicating to the public and private sectors the role of conservation banking for the various species and habitats at issue. Regional conservation goals, criteria for site selection, identification of species' needs, identification of threats to species, and other pertinent biological information should be reviewed and decisions regarding conservation banking should be made within the context of species' recovery, or in the case of non-listed species, the goal of preventing their listing. Public comment and input on the conservation strategy can be very useful.

- A conservation bank contract is required. This may take the form of a CBA (see Appendix II for a CBA template) as described in the USFWS conservation banking guidance, a BEI as described in the inter-agency wetland mitigation guidance, or an equivalent document. Because these are legally binding contracts, all parties are encouraged to get their respective legal counsel to review these agreements prior to execution. A MOA or MOU is generally not legally sufficient.
- A perpetual conservation easement grant that prohibits further development of the bank lands and restricts activities that are not consistent with the purposes of the conservation bank is required. Deed restrictions are generally not legally sufficient. The easement should be fully executed and recorded before any credit sales are approved. The easement should be held by an eligible third party (i.e. not the landowner, bank owner or USFWS). Each state has laws governing which entities may hold conservation easements, so eligible easement holders vary by state. See Appendix III for a conservation easement template.
- A long-term management plan, which has the primary goal of maintaining the habitat for the species for which the bank was established, is required. This plan should have sufficient detail to be able to determine the amount of funding required to implement effective monitoring and management of the bank lands, and sufficient flexibility to adapt the management measures accordingly when changes in conditions at the bank lands occur.
- Sufficient funding to implement the management plan is required and provisions for contingencies should be included. This funding generally takes the form of a non-wasting endowment (i.e. a fund that generates enough interest to cover management costs without depleting the principle in the fund). The fund should be managed conservatively (i.e. low risk investment), but the interest income must be sufficient to prevent inflation from depreciating the value of the fund. The long-term management endowment should be provided upfront or completely funded within a couple of years from the date the bank opens. See Chapter 9 for detailed information on management endowments.
- The bank's service area is the agency-approved area in which the bank's credits may be used to offset projects' impacts. It should be clearly identified through a map and written description. The service area should be a biologically justifiable area such as a watershed or species recovery unit in which the conservation bank is located. It is common for service areas to be constrained for other than biological reasons, like a desire by local governments to keep impacts and mitigation within their local planning area. However, within those constraints the service areas should reflect the needs of the species or other ecological resources for which the bank is to be established.
- Credits are the quantification of a species' or habitat's conservation values within a conservation bank. The credit system, which is usually based on the species'

conservation strategy if one exists, should be clearly described and easy to use in conjunction with the methodology used to assess impacts to the species or habitat. Temporal loss should be considered when awarding and releasing credits in the case of restored or created habitat. The release of these credits should be based on successful accomplishment of clearly stated, quantifiable goals. Availability of credits and debits should be tracked in a database as they occur to prevent oversales of credits.
- Establishing banks in phases is acceptable. However, the first phase of the bank must be large enough to be self-sustaining in case no further phases are realized.

A copy of the USFWS guidance is included in Appendix I.

More lessons learned

Conservation banking is still a relatively new tool and delays in bank approvals have occurred and are likely to continue to occur in some cases. The most frequent complaint of conservation bankers is the time required to get a bank approved by the agencies. Delays resulted for a variety of reasons on both the side of agencies and prospective bankers. Many early bank proposals, and even some very recent proposals, have posed novel challenges that had to be addressed before these banks could be approved. Such challenges have included:

- The title report for the proposed bank lands revealed rights to the property had been granted in the past that were in conflict with the purposes of the conservation bank. Methods of clearing these titles depend on the nature of the rights that were granted, but in most cases these problems can be resolved.
- One or more outstanding mortgages occur on the bank lands that could result in dissolution of the conservation bank if the property was foreclosed on in the future. A subordination agreement between the landowner and mortgage holder in which the mortgage holder agrees to uphold the CBA should the property default will often resolve this problem.
- Hunting on the proposed bank lands for non-listed game species is requested. It is usually possible to come to some agreement where the amount and timing of hunting activities is consistent with the purposes of the conservation bank.
- The prospective bank owner requests multiple, overlapping species credits. This can be complicated (see Chapter 7 and Chapter 11 regarding crediting), but it is possible in some cases to accommodate these requests without allowing 'double-dipping', a situation in which a credit is sold more than once.
- The property is encumbered by an existing easement that is not adequate to meet the purposes of the proposed conservation bank. The solution is usually a new conservation easement which supersedes the old easement upon execution and recording of the new easement. Termination of the old easement generally requires consent of all original signatories.

- Difficulties in adequately assessing long-term operating and management costs often occur and underestimates are the norm. This problem has been solved for many prospective bank owners by the use of a computer program, Property Analysis Record (PAR), developed by the Center for Natural Lands Management expressly for the purpose of determining both short- and long-term costs associated with establishing and managing preserves. See Chapter 9 for a discussion of financial considerations and a description of the PAR.
- The prospective bank owner cannot provide funding for the long-term management endowment upfront. This problem can generally be resolved by use of an escrow account in which an agreed upon portion of each credit sale is diverted into the endowment account. However, the length of time it takes to fully fund the endowment affects the total amount of funds that will be needed. Just a few years' delay may more than double the amount needed depending on inflation, interest rates and other economic factors.
- Conflicts may occur over service area determinations that are not easily resolved without further study. For example, demographic or genetic studies may be needed to determine a species' population structure. These conflicts are often resolved by initially designating a conservative service area with the assurance that if future studies warrant a larger service area, the bank agreement will be amended.
- Bank proposals that are linked to larger conservation efforts such as county-wide HCPs or regional programmatic consultations may experience long delays prior to approval. Often an interim strategy that allows conservation banks to be approved and begin operating ahead of completion of the HCP or consultation can be negotiated.
- If a joint conservation–wetland mitigation bank is proposed, it is usually better to proceed with the wetland mitigation banking process and use a BEI rather than a CBA. This is because the wetland mitigation banking process is more involved, and it is easier to adjust the BEI to accommodate species conservation banking issues than vice versa. Because both species and wetland resources are at stake, it is usually appropriate for USACE and USFWS/NOAA Fisheries to be co-chairs. This is because under the current guidance, the 'chair' has the authority to approve the bank.
- Even though the CBA or BEI has been executed, delays in recording the conservation easement grant will prolong the opening of a bank and its ability to sell credits. Prospective bank owners should carefully consider eligible conservation easement holders and the time they require to review and record easements.
- There may be a change in landownership during the bank approval process. This is usually not a problem, but does cause some delay.
- There may be discrepancies between the preliminary title report and the final title report. One reason for this is that landowners may have granted rights to the property or mortgaged the property during the bank review process. Such changes can significantly delay or preclude approval of the conservation bank.
- Delays occur when significant changes in the original proposal are made on the part of the prospective banker after agency reviews have occurred. For example, requests for additional credit types, inclusion of another activity on the bank

site and changes in the footprint of the bank lands can all lead to significant delays in approval.
- Government agencies may experience staff turnover during the conservation bank review process. New staff will usually need time to familiarize themselves with the project.

As new challenges are presented, the solution toolbox to address these challenges grows. Overall, the length of time required for agency review and approval for conservation banks has improved considerably. Conservation banking guidance and templates for banking documents which facilitate the process are now available to prospective bankers.

Summary

Although California has made the most extensive use of conservation banking, this tool is also being used in other states. Banks have been established in Texas for the golden-cheeked warbler (*Dendroica chrysoparia*), Houston toad (*Bufo houstonensis*) bone cave harvestman spider (*Texella reyesi*) and several other karst invertebrates; in Arizona for the Pima pineapple cactus (*Coryphantha scheeri* var. *robustispina*), in Colorado for Preble's meadow jumping mouse (*Zapus hudsonius preblei*), in Alabama for the gopher tortoise (*Gopherus polyphemus*) and in several south-eastern states for the red-cockaded woodpecker (*Picoides borealis*). By January 2007, more than 70 conservation banks had been approved in the US by USFWS or by USFWS in conjunction with other agencies, including a bank for the nightingale reed-warbler in Saipan, a commonwealth nation of the US. During this same period NOAA Fisheries has approved only one conservation bank; however, NOAA Fisheries is working with other agencies on a pending umbrella mitigation banking agreement, the first agreement of this type, for listed anadromous fish species in California's Central Valley. Figure 2.1 shows the annual and cumulative rates of establishment for conservation banks approved by USFWS and NOAA Fisheries through 2006. Several more banks have been approved by state wildlife agencies for state listed species. The USFWS-approved banks preserve in perpetuity approximately 70,000 acres of habitat for more than 50 federally listed species and many other at-risk species.

Who owns these conservation banks? Individuals, companies, corporations, utilities, governmental agencies, non-profit organizations and land trusts. Some examples include small business owners including ranchers and farmers, real estate developers, the Colorado Department of Transportation, Bank of America, a troop of Boy Scouts and a Boys and Girls club. Conservation banking is growing in popularity as more landowners become increasingly comfortable with the idea of forgoing the possibility of future development of their lands to establish permanent preserves. A few successful bank owners have launched entrepreneurial companies specializing in mitigation banking.

Although detailed assessments of the biological effectiveness of conservation

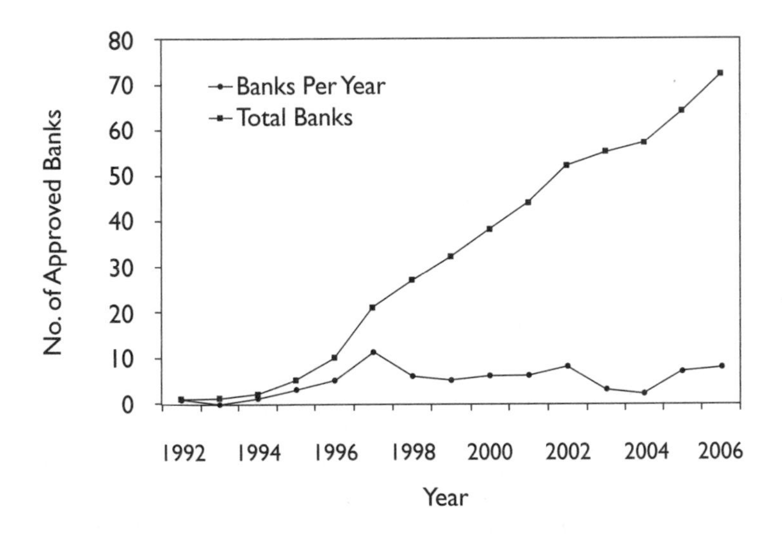

Figure 2.1 *Rates of conservation bank establishment for banks approved by the US Fish and Wildlife Service through 2006*

Source: Modified from Fox and Nino-Murcia (2005)

banking, and the comparative effectiveness of conservation banking versus project-by-project mitigation, are not available yet, the general sense is that conservation banking is working. Success can be measured, in part, by the apparent sustained quality of the habitat that most of these conservation banks protect for the listed and rare species for which they were established. Annual reports from easement holders, agency staff and bank managers are encouraging. And, in an effort to prevent bank failures, agencies have exercised their authority to suspend credit sales for conservation banks that were not in compliance with the terms and conditions of their contacts. Economic success can be measured by the number of credit transactions that have occurred and the fact that most of the banks established prior to 2000 have sold out of credits. A few bank owners have gone on to establish additional conservation banks or have added additional lands to their original banks for additional credits.

Conservation banking is likely to be most successful if it is approached as part of a larger conservation framework that includes public scoping, development of alternatives, regional planning based on the ecological needs of the species, design of sustainable preserve systems, sequencing (i.e. using a directional approach in which avoidance of impacts precedes minimization and mitigation of impacts) and the use of both on- and off-site mitigation as biologically appropriate. The future of conservation banking is likely to continue to be complex, varied and dictated by the changing conservation needs of species and economic needs of communities. Cooperative partnerships among the private and public sectors that seek long-term solutions to species conservation are essential to sustaining ecosystems and economies. Conservation banking has an important role in long-term species and habitat conservation.

References

City of San Diego (2007) 'Multiple species conservation plan', www.sandiego.gov/planning/mscp, accessed in May 2007

DeWeese, J. (1994) 'An Evaluation of Selected Wetland Creation Projects Authorized through the Corps of Engineers Section 404 Program', Report by the US Fish and Wildlife Service, May 1994, USFWS, Sacramento

Environmental Defense Fund (1999) 'Mitigation Banking as an Endangered Species Conservation Tool', Report by the Environmental Defense Fund in cooperation with Sustainable Conservation, November 1999, The Environmental Defense Fund, Washington DC

Fox, J. and Nino-Murcia, A. (2005) 'Status of species conservation banking in the United States', *Conservation Biology*, vol 19, pp996–1007

Huxel, G. R. and Hastings, A. (1999) 'Habitat loss, fragmentation, and restoration', *Restoration Ecology*, vol 7, pp309–315

Marsh, L. L., Porter, D. R. and Salvesen, D. A. (1996) 'Introduction and overview', in Marsh, L. L., Porter, D. R. and Salvesen, D. A. (eds) *Mitigation Banking: Theory and Practice*, Island Press, Washington DC, pp1–14

McElfish, J. M. Jr. and Nicholas, S. (1996) 'Structure and experience of wetland mitigation banks', in Marsh, L. L., Porter, D. R. and Salvesen, D. A. (eds) *Mitigation Banking: Theory and Practice*, Island Press, Washington DC, pp15–36

Redmond, A., Bates, T., Bernadino, F. and Rhodes, R. M. (1996) 'State mitigation banking programs: The Florida experience', in Marsh, L. L., Porter, D. R. and Salvesen, D. A. (eds) *Mitigation Banking: Theory and Practice*, Island Press, Washington DC, pp54–75

US Fish and Wildlife Service (1983) 'Interim guidance on mitigation banks', issued through a Memorandum to Regional Directors, Ecological Services Instructional Memorandum No. 80, June 1983, US Department of the Interior, Fish and Wildlife Service, Washington DC

US Fish and Wildlife Service (2003) 'Guidance for the establishment, use, and operation of conservation banks', issued through a Memorandum to Regional Directors, 2 May 2003, US Department of the Interior, Fish and Wildlife Service Washington DC

US Fish and Wildlife Service (2007) 'General statistics for endangered species', http://ecos.fws.gov/tess_public/SummaryStatistics.do, accessed 15 May 2007

Wheeler, D. P. and Strock, J. M. (1995) 'Official policy on conservation banks', issued jointly by The Resources Agency and California Environmental Protection Agency, 7 April 1995

3

The Advantages and Opportunities

Wayne White

The battle between business and biodiversity is one long fought and not easily overcome. The expansion of one and resultant depletion of the other has long caused debate among top players on both sides. The mutually agreed upon solution has taken the form of compensatory mitigation.

In the development and implementation of this new peace, however, we are finding that the business and biodiversity are not as mutually exclusive as once thought. Land developers performing their own on-site or off-site mitigation (or turn-key mitigation) has over time resulted in isolated and relatively small mitigation sites – that maintain little, if any, of the biological value that they were set aside for. On the other hand, when conservation banks are approved by the agencies involved, the resulting mitigation action on the ground has better achieved the planned biological outcomes while also maintaining larger preserve sites, and better conservation outcomes. Mitigation through the use of well-managed conservation banks has become recognized as beneficial to all parties involved in the process. Conservation banking is emerging as a practice that defies old barriers and brings biology and business into a symbiotic state, serving the interests of business people, regional and city planners, ecologists, biologists, government agencies and private interests.

Nonetheless, mitigation banks are still viewed with some uneasiness by government agency staff and some environmentalists. The purpose of this chapter is to discuss the benefits of conservation banking from both the business and the ecological perspectives. Despite some scepticism about its practice, the benefits of conservation banking clearly outweigh the current alternatives. In addition, it provides a common ground on which parties that were previously on opposing sides of an issue can surprisingly find themselves in agreement – to the benefit of all involved, as well as to the environment.

Banking broadens the conservation opportunities

To be clear, conservation banking is a practice risen from the mitigation policies created by regulatory and legislative authorities, that preserves natural habitats and ecosystems in the face of development and encroachment. At its core, it is simply mitigation – offsetting or compensating for unavoidable impacts to habitats or species. But conservation banking has evolved beyond the simple concept of mitigation for the sake of mitigating. It has become a practice that satisfies not only the laws and their requirements but goes further, and creates business practices that suit the needs of developers, private investors and natural resource agencies. Most importantly, it creates biological and ecological preserves that are environmentally worthwhile.

While traditional mitigation performed by the project proponent has resulted in ineffective, small, randomly created compensations, conservation banking is instead a collection of mitigation obligations that provides a well thought out, larger preserve with greater ecological value. Like traditional mitigation, it serves to protect appropriate ecological units in return for the development of similar ecological units elsewhere. However, conservation banking thinks beyond building in a space and saving the patch of nearby land. Instead it creates a greater land account that is drawn on to compensate for the unavoidable adverse environmental impact made elsewhere. Rather than the haphazard and geographically isolated preservation of small pockets of space typical of traditional mitigation, conservation banking allows for the preservation of larger sites chosen for their natural resource values and special-status species or sensitive habitats. By thinking ahead in planning for the identification of these sites, we are not limited strictly to the function of preserving existing natural resources, but may branch out into the restoration of former flourishing habitats, or even into the creation of new habitat spaces. Ultimately, conservation banking is mitigation to a higher power.

The primary driver in the development and operation of conservation banks is the regulatory conditions created by the agencies. In following the regulatory conditions set down by the US Fish and Wildlife Service (USFWS) and the National Marine Fisheries Service (NMFS), land developers find themselves needing to conduct due diligence in order to get approval for their projects. Enforcement of mitigation requirements for private and public sector projects will determine the market by defining mitigation needs (credit type and quantity), which will direct the development of conservation banks. Conservation bankers will begin to pinpoint the placement of banks based on the market locale (the ability to sell credits) and biological value of a site, which ultimately leads to the conservation of a species or its habitat rather than a haphazard dotting of ineffective mitigation sites (see Figures 3.1, 3.2 and 3.3). By combining the mitigation obligation of a group of developers, a larger area with greater ecological benefits may be preserved than if each developer attempted to manage their own offsets independently.

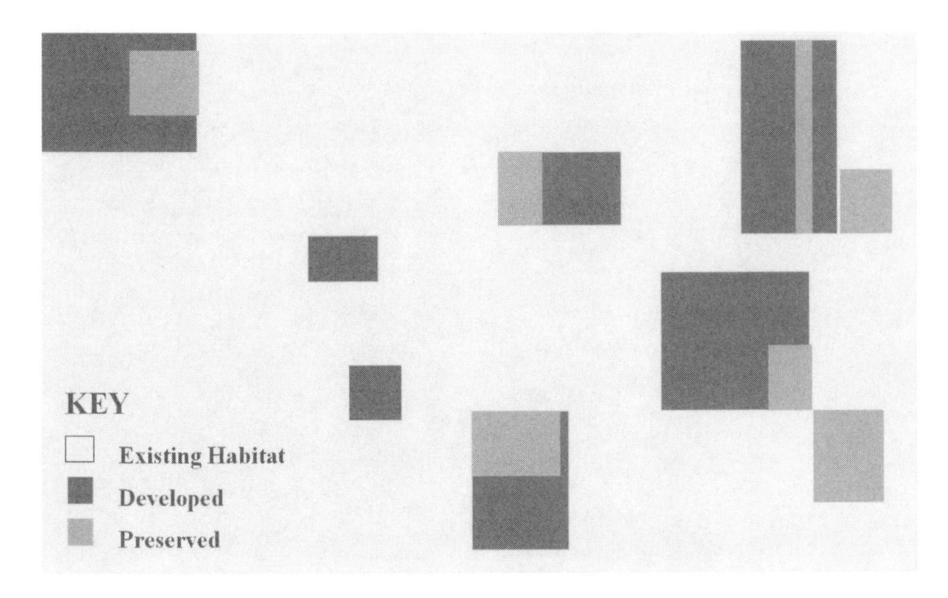

Figure 3.1 *Unplanned development in early stages*

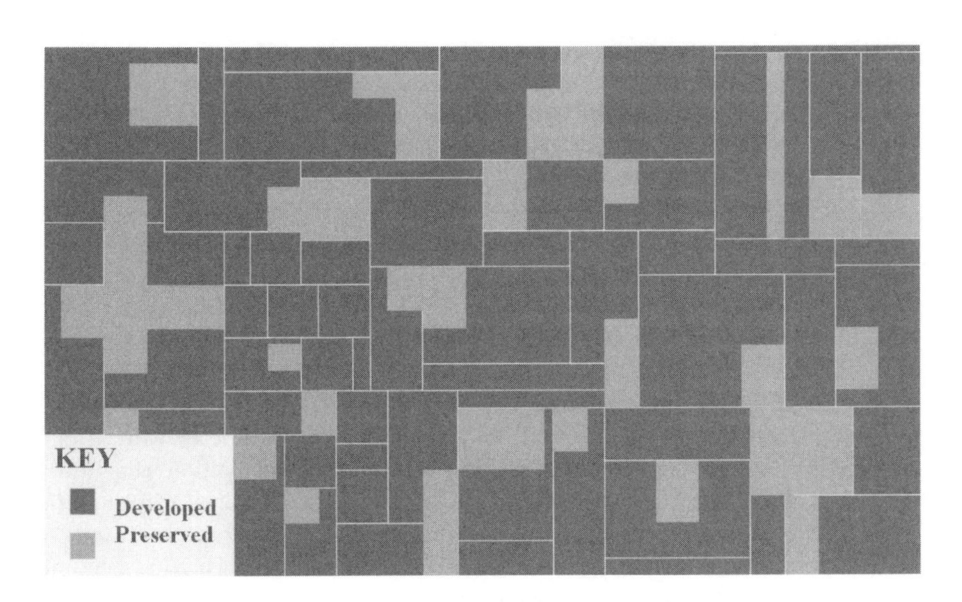

Figure 3.2 *Unplanned development when completed*

Figure 3.3 *Planned development with a conservation bank*

Business advantages

The business advantages to such a system are overlapping and mutually beneficial. Large businesses, such as real estate developers, are able to reduce costs by purchasing credits from a conservation bank rather than going to the trouble of planning and implementing their own offset. Additionally, developers are not normally set up to carry out managing their mitigation obligation in perpetuity.

In the highly regulatory environment of the Central Valley of California, there are multiple federal and state authorities with overlapping interests. In this scenario, there are listed species that occupy wetlands protected by the Clean Water Act. A development project proponent will have to create an acceptable project design that addresses both the Clean Water Act and state and federal Endangered Species Acts. These overlapping jurisdictions create a regulatory quagmire that is difficult to navigate. Additionally, negotiations will typically take longer for turn-key mitigation because the details of the on-site and off-site mitigation must also be worked out with the various agencies. In executing their own turn-key mitigation, they must address all of the aspects involved, including location, biological plan, management and operations plans, endowment funds and monitoring. On the other hand, if the project proponent chooses to buy credits from a conservation bank, aspects of acceptable mitigation are part of an approved bank and are already addressed.

In contrast to turn-key mitigation, when utilizing a conservation bank, a developer merely needs to calculate and negotiate with the agencies the amount and type of mitigation needed to offset their impact, which determines the number and type of credits to be purchased through a bank. This is because conservation banks shift the responsibility, from developer to banker, for ownership of property, science-based management plans for species and habitats, operations and maintenance plans, and adequate funding through permanent endowments. Banks also provide for fulfilling the mitigation obligation of projects that involve very small impacts or small firms that lack the capacity to develop their own mitigation to satisfy the requirements of project approvals. The project proponent can then pay for credits or acres made available by an approved bank with a service area that overlays the project, bypassing the trouble of negotiating state and federal requirements, and saving time and money through a more efficient process to reach the same goal.

Public sector advantages

In addition to private interests, public agencies are using conservation banks to facilitate public projects such as highway projects where time is money and the use of a bank saves precious time. In California, Wildlands, Inc, one of the larger banking companies, has sold an average of 200 acres of credits per year to public agencies over the last four years. Also in California, the Department of Water Resources and the Army Corps of Engineers mitigated a flood control project's impact for giant garter snakes at a conservation bank – accounting for approximately 100 acres of offset on a single project. Clearly, not only do conservation banks serve the needs of the private sector, they are also invaluable to public sector efficiency as well, saving taxpayer dollars in more efficient project implementation.

In addition to serving the needs of developers, both private and public, the conservation banking system also meets the needs of local planners at the city and county levels, as well as those of their constituents. As part of local jurisdictions, planning for community expansion requires that open space for native habitat is identified and mitigated. These areas can be developed into banks that help implement the local plans as well as provide the needed mitigation for planned urbanization. Conservation banking, then, serves the needs of local interests on multiple levels. Typically, as city and county plans implemented by the private developers move from local approval to state and federal approvals, the design can be significantly changed due to mitigation requirements. Subsequent changes to local plans by state and federal authorities are made without the benefit of local needs fully taken into consideration. If local planners adequately take into account the appropriate mitigation, or amount of native habitat open space, needed in their initial regional plans, the process runs more smoothly towards their regional planned urbanization as originally envisioned.

In the end, conservation banking allows for municipalities to create better planning by blending their mitigation requirements and native habitat open space with their housing developments, ultimately creating urbanization that strategically integrates community needs such as transportation flow, housing density, community centres and locating businesses, while consolidating mitigation away from city development or integrating open space with housing to provide for higher quality of life opportunities. In some of the fastest growing regions of California, banking allows municipalities to move ahead with reduced permitting time for projects, thus acting as an escape valve for this rapid population growth.

Private landowner advantages

Reaching beyond large business interests, and municipal interests, into the realm of the private sector, we find a new type of business formed, derived from requiring mitigation. Banking creates a for-profit, market-driven industry that involves the private sector in a non-consumptive utilization of natural resources. It allows investors who are looking to conserve, rather than consume, natural resources to endow their resources in a productive way that serves not only their own needs, but also the needs of those who would consume natural resources. Conservation banking creates a system of land stewardship that allows for the privatization of providing ecosystem services to lie in the hands of those who may otherwise have used their property for consumptive purposes to turn a profit.

This practice brings to light one of the criticisms of conservation banking, that through its efficiency, the practice promotes greater development. Granted, the practice of conservation banking does rely on development to occur, and this development does consume native habitats. However, few development projects are denied approval and they continue to march through the landscape with varying degrees of successful mitigation independent of the banking system. Rather than encouraging rampant development, conservation banking has actually developed a private business sector that is protecting habitats. This practice is even more important to preservation in light of the funding shortfalls of federal budgets and the reduced contributions of non-profit organizations that previously have been the primary mechanisms in environmental conservation.

In further developing the concept of land stewardship, private landowners also stand to benefit from conservation banking, in that they have the ability to become conservators. Conservation banks provide a carrot for landowners looking to maintain their lifestyle that in some cases goes back several generations. Both cattle ranchers who face variable prices for their beef, and younger generations that see lower economic output for ranching, wish to maintain their lifestyle and are looking to find compatible uses for their land that will allow them to do so. By partnering with conservation bankers or venturing into banking by themselves, they are then able to diversify the economic opportunities on their ranches, while preserving the land for environmental purposes. Often, banking for specific species management is compatible with minor ranch management changes, and thus make ranches good candidates for banking.

In looking further into the interests of the private sector, we see that with their participation in conservation banking, their best profit incentive may be provided by representing the best habitats, in protecting the most ecologically valuable sites. Not only, then, are our private investors profiting from preserving the environment, they are also profiting from preserving and protecting the most valuable ecosystems in a region. This trend is more apparent where the more experienced banking companies are looking to agencies for advice to focus their land acquisition efforts or review listed species recovery plans to find the most ecologically important lands to purchase and establish banks. The private sector is truly where the peace between business and environmental needs occurs. Fiscal profit and environmental preservation are not mutually exclusive in this model, as is so often the case. Here we find that the two are mutually beneficial, creating a space where both can expand together.

Ecological advantages

In finally arriving at a practice that meets the needs of business interests, it is also worth looking at the ecological advantages that conservation banking practices hold over turn-key mitigation. The intent of mitigation requirements is to avoid, minimize or, at the worst, offset impacts to the environment. Through the execution of conservation banking practices, we find the ecological benefits to be expanded beyond simply setting aside biologically unique spaces, to preserving large areas with a greater overall success at preserving habitat. As discussed, conservation banking allows for a large preserve size and consolidates compensation actions into one area, effectively eliminating postage stamp sized offsets that have little biodiversity value. This expanded preserve size then creates greater species connectivity within a protected or preserved system. This influx of species connectivity contributes to a higher level of ecosystem functionality. A criticism of current mitigation practices is that the functionality of a particular habitat is not an outcome of current mitigation offsets. In attempting to preserve larger areas, conservation banks are better at creating the ecosystem functionality currently lacking in smaller mitigation projects.

Another of the ecological advantages of conservation banking is that it can contribute to the regional strategy for species recovery. When preservation lands are strategically placed in relation to other habitats, collectively they contribute to the recovery of listed species. These lands are identified in recovery plans for listed species or Habitat Conservation Plan developed under section 10 of the Endangered Species Act. In looking deeper into the biological and ecological advantages to conservation banking, we find that banking also allows for the protection and preservation of a diverse array of habitat types beyond the intended species. The main focus of the land management may be the listed species, but the overall management also benefits the ecosystem upon which that species occurs. Because of the benefits of a conservation easement's regulatory requirements – a management plan, maintenance and operation plans, a monitoring plan to measure accomplishing biological goals, and a non-wasting endowment to fund all activity in perpetuity – conservation banking seems to be offering an ecologically superior mitigation solution.

Regulatory advantages

Beyond the business and biological advantages, the final remaining benefit of banks lies in their ability to negotiate the regulation quagmire. This benefit addresses the different regulatory mazes that conservation bankers must navigate compared to mitigation bankers to get their banking documents approved and their banks entitled. The Army Corps of Engineers, working with numerous agencies, has established the Mitigation Bank Review Team (MBRT) process that each mitigation banker must struggle through in order to get their bank approved. In some areas of the country there are records that show timelines of more than six years were needed to get a bank approved. Though this may be the extreme end of the spectrum, bankers commonly refer to approval times that are counted in years. The MBRT process is very cumbersome, with at times up to eight federal and state agencies represented on the team. Making matters worse, the agencies participate under various authorities and agency missions that at times conflict in legal interpretations when a particular agency seeks review by their in-house legal staff. Couple that with limited staffing within the agencies and minimal management oversight that allows agency staff to negotiate continuously with no deadlines, and the outcome is approval periods that often take years.

With conservation banks, the number of agencies that are required for approval is greatly reduced, often to one or two federal agencies (USFWS and/or NMFS) and the state fish and game agency. Recently, USFWS approved a conservation bank in California in less than nine months from the day the banking company began the discussions for a conservation bank with the FWS. This short approval time has not always been the case, as in the early development of conservation banking the approval times did take years, in part as a result of the agency developing agency policies and procedures; however, even then, they generally required less time than mitigation bank approvals. Policies and procedures are now better developed within USFWS due to more experience in working with bank approvals. NMFS has more recently become involved in conservation banking in the state of California, and in Washington where they are beginning to review their first fish bank. Both these federal agencies are beginning to see the benefits of conservation banks and we expect more conservation banks to be established throughout the nation where markets exist.

Some of the larger banking companies do recognize this difference in approval time and have been pursuing more conservation banks than mitigation banks in recent years. As bankers are coming to the realization of the advantages of conservation banks over mitigation banks, so are the federal and state agencies involved. Agencies are recognizing that establishing conservation banks helps implement greater recoveries for listed species when they work closely with bankers. The agencies can also influence bankers to focus on not only important recovery areas, but also those that may be under specific threat and need protection before they fall to some other non-conservation purpose.

Conclusion

In the end, with conservation banking, we come back to the bringing together of previously opposing interests in a practice that benefits all involved. Business, biology and government all have their needs met in one single practice. Development, regional planning, private interests and ecological conservation and protection, all find a common ground in a system that has risen from the regulations of government agencies, and blossomed into an entity that meets the needs of those who would develop, as well as those who would preserve, a specific natural space. Regulatory red tape is cut through a streamlining of environmental practice that is made available to those in the private as well as public sectors. Greater species and habitat recovery and preservation are made possible by the grouping of offsets from a number of sources, with a greater overall ecological impact.

Despite criticisms that conservation banking encourages development, the experience in California has been that it actually encourages more effective and efficient ecological protections, as well as providing opportunities for private landholders to use their holdings for protection rather than development. Some degree of human impact and development is an inevitable occurrence, and conservation banking is a practice that brings ecological preservation into a realm of efficiency that maximizes the conservation potential of offsetting the negative impacts of continuing developments. Ultimately, through conservation banking, we find a new peace among previously discordant groups to the benefit of all.

4

The Pitfalls and Challenges

Deborah Fleischer and Jessica Fox

The benefits of conservation banks have been detailed in the previous chapter, but banks also have their share of pitfalls and challenges. This chapter discusses the most prevalent pitfalls and challenges in conservation banking from the perspective of a range of professionals working on conservation banks, including federal and state regulators in California, ecological consultants and non-profit conservation organizations.

At the heart of the controversy is a suite of concerns around long-term management and accountability. Challenges include the potential for net loss of habitat, limited agency resources to implement and monitor banks, inadequate endowment funds to support essential land maintenance activities, competition between banks and in-lieu fees, and placement within the broader landscape, among others. And underneath these socio-political issues lies an even more basic concern: how can we be sure that banks are providing the intended benefits to the endangered species?

Gail Presley, Conservation Planning Manager with the California Department of Fish and Game (CDFG), sums it up well when she explains, 'I think banks have a lot of potential, but we also need to be cognizant of the potential for shortfalls.'

Net loss of habitat

While some conservation banks incorporate restoration activities, the majority are focused on the acquisition and preservation of existing habitat. The concern among nearly all stakeholders is that this is resulting in a net loss of habitat.

The conservation banking system is based on the working hypothesis that if you conserve large enough blocks of high quality habitat, connect it to other core preservation sites, and manage it to support the recovery of species, the species will persevere and thrive, despite the net loss of habitat.

Carol Witham, a private consultant and former President of the California Native Plant Society, believes that 'banks are fine for projects that have small, limited impacts, but when we are talking hundred of acres of land for endangered species, I think buying into banks ultimately results in more than a net loss of habitat.'

Diana Lane, a scientist with Stratus Consulting and co-author of *A Nationwide Survey of Conservation Banks*, agrees. 'When land [in the conservation bank] is not in immediate danger of development, the credits provided by the bank essentially result in a net loss of endangered species habitat' (Lane et al, 2003).

Tim Male with Environmental Defense makes the point that there is only a net loss, 'if habitat is measured in the most simplistic (and least biologically relevant) way – acres. If you instead look at the quality of the habitat and the number of acres occupied ... the species are better off.' Indeed, banking may result in higher *quality* of habitat being conserved for an individual species. Some proponents of banks argue that even with a net loss of habitat, banks do a better job protecting gene flow, avoiding gene depression, increasing breeding populations, decreasing edge effects and improving overall fitness. But the bottom line is that no scientific studies have been completed to support this conclusion.

Lack of resources to enforce and monitor

The conservation banking policy opened the door for endangered species mitigation to become an industry involving buyers, sellers, business plans, market analysts, land managers and ecologists. Unfortunately, agencies were not allocated additional staff or resources to ensure the successful implementation of this entirely new approach to mitigation. Reviewing and executing a banking agreement could easily involve an ecologist to visit the site and review the habitat management plan, an accountant to review the endowment fund estimates and maturity schedule, and a lawyer to write and review the actual agreements. The expectation that agency staff trained in wildlife biology will have the expertise to review not only the biological elements of the banks, but also the legal and financial aspects, is unrealistic. Staff have been left to work through the detailed process without adequate support from the necessary experts. Further, there has been a dearth of actual staff hours. The combination of insufficient hours and inefficient application of skills has made it difficult for agencies to effectively implement banking.

While the lack of agency resources has led to protracted agreement-processing times, the more serious ecological issue is the absence of monitoring to ensure implementation of management plans. Presley of CDFG explains: 'The agencies just are not equipped, nor funded, for the workload. There is no way we can get to all the banking sites each year to monitor them firsthand. We need to rely on banks to fulfill their obligations.' This has led to a persistent concern with conservation banking: the lack of accountability and enforcement in the system. If the agencies are not monitoring the bank sites, there is no assurance that the properties are providing the intended habitat for the species. Meanwhile, credits are being sold and impacts are taking place.

Endowment fund issues

Establishing an adequate endowment fund to support the bank property in perpetuity is critical for ensuring its long-term management. A key challenge for an endowment is to be able to incorporate all the potential needs of a property over its lifetime – you need to think about management in perpetuity. 'One of the great difficulties in developing thorough management plans and endowments is trying to identify contingencies for anything that could happen – there are lots of educated guesses', says Matt Gause, Senior Ecologist at Westervelt Ecological Services.

An article in Environmental Law Review details the potential risks of wetland mitigation banks and raises concerns over issues related to the vitality of financial assurances and the ability of regulators to take enforcement actions against a mitigation banker (Gardner and Radwan, 2005). Many of the concerns in this article also apply to conservation banks. The article points to several banks where the long-term endowment accounts are currently underfunded, raising concerns about whether the long-term stewardship needs of the sites can be met.

Oversight of endowments is also a tricky issue. Third-party mitigation landholders like the control and flexibility that controlling an endowment provides, and bankers like the idea because, typically, third-party organizations can generate a higher rate of return, requiring a smaller upfront investment.

'The State (currently) only gets 1.7 per cent return of investments because it has to go into the State Pooled Investment Fund and that doesn't even match inflation', explains private consultant Michael McCollum. 'Plus', he continues, 'if a non-profit wants to manage the property it can take a year or more to get paid for maintenance. It just doesn't work.' However, the case of The Environmental Trust (TET), a non-profit banker that went bankrupt, raises a series of concerns around adequate oversight of endowments. While some argue that a third-party organization can manage and invest the endowment more effectively than a governmental agency, TET's poor financial management practices raises the question of what accountability mechanisms must be in place to ensure an endowment is being properly managed.

Presley of CDFG explains her perspective: 'TET dipped into the principal and spent it for operations. They were operating outside of any kind of agreement with the wildlife agencies. And 4000 acres of land now has not enough money to cover it. The trust is gone and they are looking for people to take the land with very little management funding. It is a mess. It is our main reason for suggesting that the Department should continue to hold endowments even though we earn lower interest rates (the 30 year average has been 8 per cent)... Because the money is completely protected, it will never be mismanaged or misspent.'

McCollum also expresses: 'We all knew that TET was going to go sour because it was a fly-by-night operation, undercutting legitimate managers.' He expresses frustration that 'banking detractors are using TET as the poster child bogeyman to make their case against banking. However, such failures have occurred in other types of conservation and mitigation where planning and management was inadequate.'

There clearly is a need to balance the desire of regulators to protect the principal over the long term and bankers' need for control over funds to adequately steward properties. And, while many are concerned about endowments being under-funded, the flip side of the issue is the concern that the large, upfront financial requirements can make it difficult for small landowners and ranchers to get involved in a bank. For more on the issue of endowments see Chapter 9.

Competition with in-lieu fees

Prior to conservation banking, project proponents could offset their impacts by implementing their own mitigation project, or paying an in-lieu fee to the agencies. In the case of in-lieu fees, the agency is tasked with the acquisition and management of habitat lands that offset the original impacts. Now with conservation banking as an option, project proponents can purchase bank credits as mitigation. When project proponents are reviewing their options, they consider the costs and benefits of their three primary mitigation strategies: implement their own offset project; pay an in-lieu fee; or buy credits from a conservation bank.

In California, some counties have developed a fee-based system that has not kept up with the rapid increase in land values. If land prices are rapidly rising, and fees are not large enough to meet acquisition and management goals, there is no way the agencies will be able to afford to buy enough land to offset impacts, resulting in a loss of habitat. Additionally there can be a substantial time delay between the time of habitat impact and habitat mitigation.

In regions where an in-lieu fee is below market rate, bankers will not create a bank because they cannot compete economically. Royal Gardner, Professor of Law and Director of the Institute for Biodiversity Law and Policy at Stetson University College of Law, recently wrote:

> *Mitigation bankers generally view in-lieu fees with suspicion: they are not held to the same standards as mitigation banks. Mitigation bankers must invest and meet performance standards before credits can be released for sale, but in-lieu fee credits can be sold before a mitigation site is even identified. Yet mitigation banks and in-lieu fees compete for the same mitigation dollars, and in-lieu fees may be a cheaper alternative for the permittee (in part because the credit price may not reflect the true costs of future mitigation). Thus, the presence of in-lieu fees in a market can undercut demand for mitigation credits from a bank. (Gardner 2007)*

In-lieu fees must take into account true costs of land acquisition and restoration and be held to the same standards as banks.

In November 2003, Congress ordered the US Army Corps of Engineers (USACE) to establish regulations that would require the various mitigation options to meet the same standards. USACE and US Environmental Protection Agency issued draft regulations for public comment in March 2006. The draft rule proposed to phase out in-lieu fees over a five-year period, at the end of which the in-lieu fee projects would have to meet the same standards as mitigation banks. The final rule is still pending. It is important that fees are set at an appropriate market rate so they will in fact constitute legitimate mitigation and compete on a level playing field.

Placement within the landscape and adaptive management

To ensure suitability and viability of habitat, a bank requires proper placement within a broader landscape that takes into account recovery goals, genetic diversity needs, species biology, distribution and connectivity requirements, and the expected development footprint over the next 25 to 50 years and beyond. In reality, banks are often located on land that a banker already owns and without a larger conservation planning strategy. A property may ultimately be surrounded by urban development, creating a large, isolated island with limited connectivity and movement corridors. Presley of CDFG explains: 'The trick to having banks that are scattered across the landscape in isolated places is whether eventually they will be connected to something else. Or will land practices occur on adjacent lands that over a long period of time will erode the benefits of the mitigation bank.'

Maintaining a bank within a changing surrounding ecosystem will require adaptive management. Over the long term, we need a system that will allow managers to respond to changes as the ecosystem changes and the science is better understood. When asked about adaptive management in the field, McCollum responds: 'There is some minimal language in the agreements that talks about adaptive management, but in practice, we find that it is often inadequate. The lack of focus on adaptive management sets up opportunities for conflicts between the manager and oversight agencies. If problems develop, there can be strong resistance to manage differently, depending on the agency staff person.' Making things even more difficult is the reality that most agencies don't have the resources to enforce management plans, much less oversee an adaptive management system.

In response to the Endangered Species Act (ESA), and other species-specific regulations, conservation banks often focus on the habitat needs of a single species rather than the needs of the larger ecosystem. What might be good for one species might not be good for the ecosystem.

In the management plans she has seen, Witham explains that 'people are doing reactive and often single-species management without thinking about the impacts it might have on the ecosystem and without doing small studies in advance to determine the impact. Most of the management plans I have seen don't focus individual actions into ecosystem integrity and function. If you manage for one species, you can get cascading disasters (where you react to one management mistake with another mistake).'

Another concern is management practices that force a banker to artificially arrest the natural succession of a habitat in order to maintain the ideal conditions for one particular species, despite the ecosystem's natural desire to mature and go through its normal successional stages.

Lane explains: 'In general, you want your land to be in various kinds of mosaics. You want early, mid and late-successional habitats in different places because different species need each of them.'

Challenges with long-term protection

This brings us to the 'in perpetuity' question. On a legal level, in perpetuity is forever. However, how this plays out in reality can have its challenges. As land changes ownership, ensuring that easement restrictions are understood and implemented by new owners can be a challenge. In addition, keeping track of all the banking agreements and enforcing the associated conservation easements is a large task that can overwhelm land management staff.

Ecological uncertainty

To date, there has been no comprehensive investigation into the success or failure of banking from the perspective of the endangered species. The intention of habitat banking is to help reduce the influence of habitat fragmentation and offset habitat loss (US Fish and Wildlife Service (USFWS), 2003), but we don't know if the current implementation of banking is achieving this. Specific concerns include the possible net loss of habitat and the use of preserved land, the lack of adaptive management approaches, and the spatial location of bank properties within the landscape.

There is a clear lack of peer-reviewed science that looks at the ecological performance of conservation banks and their role in the recovery of endangered species. Nobody is able to answer with certainty whether conservation banks offset endangered species impacts, or if banks will function ecologically over the long term. The research to answer these questions is lacking and it is not clear whether banks are providing the intended mitigation, or whether they are doing so with more success than other mitigation options.

Conclusion

Many of the issues discussed in this chapter are already being addressed. For example, the issue with in-lieu fees is awaiting a final rule from USACE, concern over the protracted process to get banks approved is being addressed by USFWS, and discussions over who holds the endowment funds are active. The uncertainty regarding the contribution of banking to species persistence needs attention. And there are indications that future resources will be dedicated to an assessment of the ecological effectiveness of conservation banking in relation to the protection of endangered species.

Despite the pitfalls and challenges across regulatory, business and ecological fronts, Ken Sanchez, Assistant Field Supervisor of USFWS summarizes it nicely: 'With the current administration's focus on incentives or market-based solutions, mitigation banks are clearly a bright light in the political process of administering the ESA.'

And the success of an individual conservation bank appears to be greatly dependent on the execution. If located within a landscape planning context, with a biologically sound and well-funded management plan, clear lines of accountability, and enforcement, a conservation bank will maximize its chances for success.

Acknowledgements

Many thanks to the generous and busy professionals who took time to speak with us about the real-life challenges of implementing conservation banks, including: Carol Witham, Scott Wilson, Ken Sanchez, Brian Boroski, Craig Bailey, Craig Denisoff, Sean Smallwood, Diana Lane, Carl Wilcox, Gail Presley, Michael McCollum, Matt Gause, Tim Male and David Mills.

References

Gardner, R. 'Reconsidering in-lieu fees: A modest proposal', published 9 July 2007 at The Ecosystem Marketplace: http://ecosystemmarketplace.com/pages/article.opinion.php?component_id=5073&component_version_id=7494&language_id=12)

Gardner, R. C. and Radwan, P. T. J., (2005) 'What happens when a wetland mitigation bank goes bankrupt?' *ELR News & Analysis*, vol 35, pp10590–10604

Lane, D., Mills, D. and Chapman, D. (2003) A *Nationwide Survey of Conservation Banks.* Northwest Fisheries Science Center, NOAA Fisheries, Seattle, WA, www.st.nmfs.gov/st5/abstracts/A_Nationwide_Survey_of_Conservation_Banks.htm

USFWS (2003) 'Guidance for the establishment, use and operation of conservation banks', US Department of the Interior, Fish and Wildlife Service, Washington DC

Part II

Establishing A Conservation Bank

5

Ecological Considerations

Robert Bonnie and David S. Wilcove

Introduction

Six years ago, residential development in parts of Mobile County, Alabama, ground to a halt thanks to the presence of the threatened gopher tortoise and the Endangered Species Act (ESA). The county had ceased issuing permits for new septic systems on properties where gopher tortoises were present. Developers and landowners were furious. At the same time, the gopher tortoise population in the county was disappearing due to fragmentation and degradation of native longleaf pine habitat through conversion and neglect. The US Fish and Wildlife Service (USFWS) had little to offer in the way of practical solutions for landowners or tortoises.

Along came the Mobile Area Water and Sewer System (MAWSS), which owned longleaf pine forest with a diminishing though restorable population of gopher tortoises. Under a conservation banking agreement signed in June 2001, MAWSS set aside 220 acres of habitat as a gopher tortoise conservation bank from which local landowners could purchase credits for each gopher tortoise harmed by development. In exchange, MAWSS and USFWS agreed to relocate tortoises to the bank, restore habitat for the species, and permanently protect and manage the habitat – all pursuant to an ESA-approved habitat conservation plan that established the bank.

Since 2001, over 60 credits have been sold by the bank, allowing developers and landowners to develop lands formerly occupied by tortoises. The tortoises also appear to be benefiting. A monitoring programme at the bank site has documented successful reproduction by the tortoises for the first time in years. Moreover, MAWSS is considering expanding the bank significantly. Doing so would likely secure a large, viable population of tortoises for years to come.

While the Mobile conservation bank is still relatively young, its accomplishments thus far demonstrate the promise of conservation banking. Originally developed in the context of wetlands mitigation under the Clean Water Act, banking remains a controversial topic in environmental circles (Bean and Dwyer, 2000; Bauer et al, 2004; Fox et al, 2006). In theory, it holds the promise of providing greater flexibility and efficiency to regulated interests, while producing better results for the environment. Yet wetlands mitigation banking, which operates under similar principles, has been fraught with controversy. Critics contend that the practice reduces pressure on developers to avoid harm to existing wetlands while yielding restored wetlands that fulfil few of the ecological functions of the destroyed ones (Zedler, 1986; Roberts, 1993). The goal of endangered species banking is inherently simpler – to mitigate harm to a particular species, rather than restore an amorphous set of ecological functions – but the issues are complex nonetheless.

Whether conservation banking lives up to its promise will depend on whether banks are designed in an ecologically sound manner that advances the recovery of endangered species as opposed to simply providing a means for regulated interests to address their legal obligations under the ESA cheaply and quickly. In this chapter, we focus on some of the ecological issues related to designing conservation banks. We do so largely by examining two species conservation banks, including the one in Mobile for the gopher tortoise and a second bank in southern Utah for the Utah prairie dog (UPD) – both of which we had a hand in developing. We briefly describe the banks and then focus on how several key scientific issues were addressed in them. Notably, the two examples we address in this chapter involve restoration of habitat at a bank site, resulting in no net loss of habitat. In fact, in both cases, there has been a net gain of habitat as a result of credit sales. Most conservation banks today, however, preserve and manage existing habitat that is already occupied by endangered species. Still, the issues we address are not unique to the two banks that we describe, although in some cases, the solutions may be. Nor is our review of these issues intended to be exhaustive. Rather, we address a subset of the issues that are likely to be faced by anyone involved in creating, managing or regulating endangered species conservation banks.

The UPD conservation bank

Like the Mobile bank, described above, the UPD conservation bank grew out of controversy. UPDs have suffered from over a century of persecution. Landowners and government officials sought to reduce their numbers, holding the prairie dogs responsible for damage to crops, forage and irrigation systems, and believing their burrows to be a menace to cattle, horses and farm machinery. They succeeded in doing so all too well. By one estimate, the total population of Utah prairie dogs dropped from roughly 95,000 in 1920 to 3300 by 1972 (USFWS, 1991), and the species was added to the endangered species list in 1973. Since then, its numbers have rebounded somewhat, but the species is far from secure.

In its 1991 recovery plan for the Utah prairie dog, USFWS identified three main populations, separated geographically and differing with respect to vegetation: the West Desert, primarily within Iron County; the Paunsaugunt region, most of which falls within western Garfield County; and the Awapa Plateau, which sprawls across parts of Garfield, Wayne, Sevier and Piute counties.

Development pressures around Cedar City, Iron County, in occupied prairie dog habitat spawned interest in conservation banking. USFWS had been allowing private lands to be developed in exchange for habitat restoration and prairie dog relocation on public lands; the Service apparently did not see much of a future for Utah prairie dogs on private lands. Yet, despite the prevalence of public lands throughout Utah, most Utah prairie dogs (approximately 76 per cent) are found on state or private lands, not federal lands. Simply writing off conservation of habitat on non-federal lands was not a viable conservation strategy; it could not lead to a recovered prairie dog population.

The State Institutional Trust Lands Administration (SITLA), which manages some 3.7 million acres to generate revenue for Utah's public schools, saw an opportunity to increase the value of some of its less productive lands by selling endangered species credits. In particular, SITLA manages land on the Awapa Plateau that comprises virtually the entirety of one of the three designated recovery populations for the Utah prairie dog. Unfortunately, the prairie dog population on the plateau has declined greatly in recent decades and shows little sign of improvement. Remote and largely undeveloped, with little in the way of tourist infrastructure, valuable minerals or productive farm- and ranch-land, the Awapa Plateau is hardly an economic engine for the State of Utah. SITLA, however, realized that it could earn revenue from prairie dog conservation in these otherwise isolated lands if it could market conservation credits.

In September 2005, after a long period of negotiation, a conservation bank was established on three sites totalling about 800 acres on the Awapa Plateau. SITLA has placed these sites under perpetual conservation easements and established an endowment fund to underwrite management activities beneficial to prairie dogs, to be carried out by the Utah Division of Wildlife Resources. SITLA earns credits in two ways. First, it received some credits for protecting and managing an existing, albeit much diminished population at one of the sites. Second, SITLA is able to earn additional credits by restoring UPD populations at the two sites that currently lack them or by increasing the population at the site where they still occur.

The bank opened with 77 credits to sell. Rapidly expanding Iron County immediately purchased all the credits and intends to resell them to developers. The selling price was $1636 each, plus $200 per credit for the perpetual endowment fund, a price determined by a third party that appraised the sites. As habitat restoration generates more credits, SITLA can sell those credits or use them to mitigate its own development projects elsewhere in the range of the Utah prairie dog. SITLA also has the option of expanding the bank to generate still more credits.

With this background on the gopher tortoise and UPD conservation banks, we now turn to some of the key ecological issues that had to be addressed in each case.

Ecological considerations of conservation banking

In a legal sense, the success or failure of a conservation bank is determined by whether it fulfils the requirements for mitigation set forth in the ESA. In the case of banks that are intended to mitigate for habitat losses on private lands, the ESA requires that the person or institution engages in the take of a listed species 'to the maximum extent practicable, minimize and mitigate the impacts of such taking', and that 'the taking will not appreciably reduce the likelihood of the survival and recovery of the species in the wild…' In these instances, the ESA requires a written habitat conservation plan (HCP) outlining how mitigation will be carried out. In many instances, it may be necessary for the owner of the conservation bank to also develop a conservation banking agreement which spells out how credits will be sold and what conservation activities will be undertaken at the bank. Some conservation banks provide mitigation for habitat losses due to activities carried out, funded or permitted by the federal government. In this case, the ESA requires that such activities 'not jeopardize the continued existence' of the species, and mitigation may then be required through an incidental take statement from USFWS. No HCP is required in such instances, but the bank owner may need to develop a conservation banking agreement. In either case, fulfilling the requirements of the ESA via the creation of a conservation bank requires significant scientific analysis. Below, we discuss nine key issues.

Species suitability for banking

Perhaps the first question to ask regarding conservation banking is whether it is suitable for all species. The answer, we believe, is no. For example, establishment of conservation banks for wide-ranging species, such as grizzly bears or Florida panthers, seems highly impractical due to the vast amount of habitat needed to sustain breeding populations of such species, and the remote likelihood that private banks can provide sufficient land. The home range of a single grizzly bear, for example, can cover 30–160 square miles. Likewise, conservation banking for anadromous fish or other wide-ranging aquatic species presents problems since such species are vulnerable to habitat loss throughout entire river systems and watersheds. Transferring that loss of habitat from one place to another within a river basin is unlikely to result in bona fide benefits to the species in question.

Moreover, since conservation banks typically involve the management and/or restoration of habitat, a lack of knowledge of the habitat needs of a particular species can render that animal unsuitable for banking. If there is significant uncertainty about the life history, habitat preference and/or management of a species, then conservation banking becomes very difficult, if not impossible. Banks require establishment of habitat guidelines and population goals, the determination of mitigation ratios, effective management and monitoring – all of which can be difficult even for species that are well understood. The phrase 'if you build it, they will come' loses meaning when no one knows what the 'it' is.

That said, there is always scientific uncertainty about the ecology of a given species (especially so with endangered species). In the case of the gopher tortoise, when the Mobile conservation bank was established, scientists had not yet precisely quantified the amount of habitat required for breeding and foraging by individual gopher tortoises. While the recovery plan provided some guidance, neither it nor any other published document could answer the critical question of how much habitat per tortoise should be protected and managed at the bank. To obtain at least a provisional answer to this question, a number of prominent scientists were consulted as the bank was being developed.

An understanding of the most important threats facing a species is also important.

For example, populations of Utah prairie dogs, like all prairie dogs, periodically are wiped out by sylvatic plague. Deliberate isolation of discrete populations, therefore, might be part of a comprehensive recovery strategy – an important consideration in siting conservation banks. Of course, if the populations are isolated from each other, then care must be taken to make each population large enough to maintain itself over time without the benefit of immigration from outside.

Knowledge of the dispersal capabilities of the species in question is also useful for determining whether colonists are likely to interact with species on preserved habitats or discover restored habitat at conservation banks. For banks that rely on preservation of existing habitats, interaction with populations on other lands may be critical for maintaining a viable population at the bank site. Dispersal is also an issue for banks that are created by restoring habitat. Is the species in question likely to colonize the restored habitat on its own, or will it need to be translocated to the site? If the latter, then the availability of 'extra' individuals for translocation and the ability of wildlife officials to safely translocate them become critical issues. In the case of the endangered red-cockaded woodpecker, which inhabits fire-maintained pine forests in the southern coastal plain and piedmont, the improvement of translocation methods, through provisioning of artificial nest cavities and movement of juvenile birds, has made conservation banking much more feasible because new populations can be established rapidly.

In the case of the Utah prairie dog, early translocation efforts met with little success, but over the past decade, improved techniques, including supplemental feeding and water, use of retention cages and predator control, have boosted the success rate. Nonetheless, even with these improvements, translocation of Utah prairie dogs remains highly labour intensive, increasing the cost of creating conservation banks.

Where banks involve restoration of habitat, perhaps the most important factor in determining the suitability of a species for conservation banking is simply whether its habitat can be restored in a reasonable time frame. We suspect that a disproportionate share of conservation banks that focus on restoring habitats in the future will be created for early-successional species because habitats for such species may be among the easiest to restore.

Restoration versus maintenance

Most endangered species find their way onto the endangered species list due to habitat loss (Wilcove et al, 1998). It follows, then, that rebuilding their populations will require restoration of habitat. Ideally, conservation banking would result in no net loss of habitat and, preferably, would actually increase available habitat over time through mitigation ratios that require more habitat to be restored than taken (see below, 'Currency and mitigation ratios').

However, there may be instances when allowing take of an endangered species in exchange for permanent protection of existing, occupied habitat is acceptable. Wilcove and Chen (1998) found that at least 60 per cent of US endangered species were imperilled by invasive species or fire suppression. For such species, it is necessary to both protect the existing habitat against development, and also actively manage the site. Although the ESA prevents private landowners from undertaking actions that harm listed species, it does not force them to maintain their property in a condition suitable for listed species. In other words, it cannot compel a landowner to burn a portion of her property to maintain a particular successional stage or to root out an invasive plant that is out-competing the native vegetation (Bonnie, 1999). Under such circumstances, it is conceivable that a listed species would be better off with a smaller amount of habitat that is properly managed compared with a larger amount that is neglected. If so, then the creation of a conservation bank based on the long-term management of currently suitable habitat (via an endowment fund, for example) would be worth pursuing. Indeed, the majority of conservation banks created thus far preserve pre-existing habitat. A primary rationale for these banks is that they guarantee both the protection and proper management of existing habitat and, in so doing, contribute to the long-term viability of the species in question.

One could also imagine a banking arrangement that exchanges small, isolated, remnant populations of species for permanent protection and management of strategically located parcels of prime, occupied habitat. In raising these issues, we do not mean to endorse categorically habitat-maintenance banks (just as we wouldn't categorically endorse banks that produce credits through restoration). Rather, we think the issue is one that should be considered on a case-by-case basis.

In Mobile, the conservation bank already had a small population of about ten tortoises. Establishment of the bank will maintain their existing habitat and restore enough additional habitat to establish a viable population. Likewise, in Utah, the Awapa Plateau contained a small number of prairie dogs prior to establishment of the bank. The bank will contribute to the recovery of the UPD both by protecting and managing its habitat and by restoring habitat for a larger population.

Currently, most mitigation banks generate credits largely through the preservation of occupied habitat. However, even in these instances, habitat management and restoration activities can boost populations. Given the importance of habitat management and restoration to the conservation of endangered species, we believe that in most cases banking will involve a combination of preservation and restoration activities. What is undeniably true in all cases is that the bank must have an endowment or other source of ongoing funding sufficient to cover all of the management activities needed to sustain the species in question.

Suitability of conservation banking sites

A key question in judging the ecological benefits of a proposed conservation bank is the suitability of the bank site. Typically, larger bank sites are preferable to small ones, and sites contiguous to other protected lands are preferable to isolated ones. However, smaller bank sites that serve to link larger populations could be particularly beneficial to recovery efforts. Similarly, banks can be a means to build up a small, faltering population that is currently not viable but is considered important to the long-term recovery of the species. This is exactly the case with the UPD conservation bank on the Awapa Plateau.

A further consideration in judging the value of a proposed bank is whether the bank owners are likely to be able to properly manage the species at the site. For example, proximity to urban populations might restrict the use of prescribed fire on bank sites where fire is an important tool for maintaining suitable habitat. The attitudes of neighbouring landowners should also be considered, especially in regions where there is strong resistance to endangered species conservation. Safe Harbor agreements and other tools can be used to provide assurances to those neighbours that the presence of a bank in the neighbourhood will not increase legal restrictions on nearby properties.

Finally, a growing concern in siting conservation banks is the effect of climate change on the persistence of populations at the site. This may be a particularly important consideration for banks situated near the southern or upper altitudinal limits of a species' range. Drought, flooding and water-availability issues related to climate change should also be considered. These are probably best tackled on a case-by-case basis given the difficulty in predicting the magnitude and impact of climate changes in specific regions.

Service area

A key consideration in designing a conservation bank is determining the 'service area' that can be served by the bank. For example, in the case of a species with a large range, is it appropriate to use credits from a conservation bank in one state or region to offset habitat loss occurring in another?

In wetlands mitigation banking, the service area of mitigation banks is an especially important question because many of the services that wetlands provide, such as water filtration or flood control, are locally important. When wetlands are destroyed, a strong argument can be made that mitigation must occur within the same watershed so as to compensate for the loss of local services (Bean and Dwyer 2000).

The case with endangered species is different. While endangered-species' habitat may provide aesthetic or other benefits to a particular locale, the goal of conservation efforts is to preserve enough viable populations so as to assure the continued existence of the species. And while the geographic distribution of species is important to preserve genetic diversity and to insure against catastrophic losses due to disease or natural disasters, the particular locale of any one population is less important. Thus, one can argue that for many species, it may be entirely defensible to allow take of habitat in one region in exchange for mitigation in another, assuming the conservation of genetic resources.

In the case of the Mobile gopher tortoise conservation bank, the bank service area was limited to Mobile County. This decision was not made on biological grounds. Instead, Mobile County was one of the few areas where enforcement of the ESA with respect to gopher tortoises was actively occurring. In fact, the County government was enforcing the Act by refusing to issue septic-system permits for new homes on tracts where active tortoise burrows were found, for fear of running afoul of the law. As a result, demand for incidental-take permits that would allow development in gopher tortoise habitat was largely confined to Mobile. Perhaps more important, the Mobile Area Water and Sewer System (MAWSS), which owned the bank site, was a public utility governed by commissioners appointed by local politicians. There was a strong incentive for MAWSS to be responsive to local problems. (The need to provide local mitigation to landowners and developers probably also encouraged the MAWSS commissioners to set a reasonable price for mitigation credits.)

SITLA's conservation bank on the Awapa Plateau is a different story. In this case, the demand for conservation credits came largely from developers and landowners around Cedar City, Utah, nearly at the south-western corner of the Utah prairie dog's range. The Awapa Plateau on the other hand, is located in the north-eastern quadrant of the species' range. However, because the Awapa Plateau is designated as one of three recovery areas, its significance to the recovery of the UPD is immense. USFWS, therefore, determined that a large service territory was warranted, and granted SITLA permission to engage in trades throughout the range of the prairie dog.

Determination of a bank's service area has important implications for the economics of conservation banking. A narrowly proscribed service area may limit the demand for credits and could make conservation banking less financially attractive for the owner of the bank. In the case of the UPD, land values are much lower on the Awapa Plateau than near Cedar City. This difference in land values makes conservation banking financially attractive both for developers and SITLA (for example, see Bonnie, 2004), provided they can do so across the two regions. Had USFWS restricted the service area of the bank to only the Awapa Plateau, trades between Cedar City and the Awapa Plateau would not have been possible, and the bank would have been an economic bust. A counterbalancing consideration, however, is to ensure that there are also ample incentives available to landowners to conserve UPDs around Cedar City in order to ensure the continued existence of UPDs in that portion of the species' range. While economic considerations might seem out of place in a chapter about ecological issues related to conservation banking, if the economics of banking aren't viable for both seller and purchaser, the ecological questions become irrelevant.

Currency and mitigation ratios

One of the more challenging issues in designing conservation banks is determining the currency that will be traded. In other words, how does one translate the amount of habitat that is destroyed or the number of individuals of a listed species that are killed ('take' in the parlance of the ESA) into an amount of habitat or number of individuals that will be protected or restored at the bank site? The most straightforward approach is to use acres of habitat as the currency for mitigation, which is what most banks do (Fox and Nino-Murcia, 2005). Thus, for example, destruction of ten acres of habitat would require that ten acres of habitat be protected or restored at the bank. For most species, acres of habitat makes sense, but it is often a crude measure of the impact of habitat destruction on species' populations.

Take the case of the red-cockaded woodpecker, which is a territorial, cooperatively breeding bird that constructs its cavities in living pines. These cavities are typically arrayed in 'clusters' and the woodpeckers forage in the surrounding mature pine forest. Destruction of 200 acres of occupied habitat could result in the loss of 4–5 breeding pairs if the habitat is of particularly high quality, or it could result in the loss of only one breeding pair if the habitat is marginal. Likewise, at the conservation bank, protection of 200 acres of habitat does not provide a reliable picture of the value of those mitigation activities to the woodpecker because the acres do not necessarily equate to the bird's population. Thus, using simple acreage as the currency is often not a good measure of the actual impacts of habitat destruction nor of the mitigation activities undertaken at the bank.

Because the woodpeckers live in discrete breeding groups, developing a 'currency' other than acres of occupied habitat should be quite straightforward. Because woodpeckers live in discrete breeding groups, a logical solution would be to use the number of breeding groups as the currency. Yet, even in this case, one must decide how much habitat is required to be protected and managed for each breeding group at the conservation bank in order to be confident that the bank is truly mitigating for the loss of breeding pairs in the service area. The USFWS recovery plan for the red-cockaded woodpecker defines minimum standards for woodpecker breeding and foraging habitat. Thus, the currency might be better defined as breeding groups of woodpeckers with enough surrounding, suitable habitat at the conservation bank to meet the minimum guidelines in the recovery plan. (This, of course, raises the question as to whether the guidelines in the recovery plan are sufficient to ensure healthy populations of this endangered species – a question we will not address here, but which is nonetheless an important consideration in designing a currency for conservation banking.)

In the case of the gopher tortoise, the MAWSS bank translates the number of animals taken at the development site – in this case, tortoises, not breeding pairs – into a measure of occupied habitat at the conservation bank. Because there were no habitat guidelines in the gopher tortoise recovery plan, bank planners estimated the total population of tortoises that could be sustained at the bank (i.e. the bank's probable carrying capacity) by drawing upon the existing scientific literature and the opinions of tortoise experts. One option would have been to simply take this population number and subtract the existing number of tortoises at the bank to calculate the number of credits that could be sold. This, however, would not allow natural growth of the population because the target density of tortoises would have been reached through translocation of tortoises to the site once all credits were sold. So, instead, USFWS and MAWSS chose to base the number of credits that could be sold on 80 per cent of the calculated carrying capacity. This had two effects. First, it allowed for natural growth of the population because it meant that there was additional habitat that could be occupied by juvenile tortoises. In addition, reducing the number of credits to be sold increased the amount of habitat that would be protected for each tortoise harmed at the site where development was to occur.

UPD population tallies are based on the number of adults estimated by visual surveys conducted in the spring. Because UPD populations can fluctuate significantly from year to year, regulators and bank managers use a multi-step process to determine how many UPDs are at a given site (for the purposes of determining credits). First, the bank takes the highest population count from the previous five years of spring surveys as its tally of adult UPDs at a given site. It then multiplies that value by a factor greater than 1.0 to account for juvenile prairie dogs (which are typically underground when the surveys are taken). The resulting number is then used as the estimate of the number of UPDs in a given locale. USFWS requires that two credits be purchased for each prairie dog taken at the development site. Each credit corresponds to one acre of protected, managed habitat at the conservation bank. Thus, like the gopher tortoise bank, the UPD bank translates the number of individuals taken into a number of acres of occupied habitat that must be protected and managed at the bank.

In the case of the woodpecker, prairie dog and tortoise, surveying individuals is relatively straightforward, and biologists can produce accurate estimates of individuals (or, in the case of the woodpecker, breeding groups). Determining precise numbers for other listed species is more difficult. For example, the flatwoods salamander, like the woodpecker and tortoise, is a resident of the longleaf pine ecosystem, but it spends a significant portion of its life underground, making surveys very difficult. In a case like this, the currency should probably be based on the number of acres of occupied habitat that are taken instead of any estimate of the number of individuals taken. A similar situation might apply in the case of species whose populations fluctuate significantly from year to year. Thus, the number of acres of Karner blue butterfly habitat that are destroyed by a given activity may be a more meaningful measure of that activity's impact than the number of butterflies killed; acres rather than individuals would be the best currency.

In an ideal world, a more precise approach to determining the currency would be to quantify the impact of the loss of habitat on the viability of the species in question. One would then calculate the appropriate amount of habitat to restore at the bank in order to ensure that the trade did not diminish the species' prospects for survival. Thus, if the loss of habitat lowered the survival prospects of an endangered species by 2 per cent, then mitigation activities should be required to boost the survival prospects by at least 2 per cent. Such a system would take into account the spatial configuration and quality of the habitat and the population size and structure of the species in question. The rapidly maturing field of population viability analysis (see Groom et al, 2006) is devoted in large part to just these sorts of calculations. However, population viability analyses typically require data on the fecundity, survivorship and dispersal abilities of the species across a range of habitat types, and these data can be very difficult and expensive to obtain. In a study examining the status of 36 imperilled plants and animals in central Florida, Turner et al (2006) concluded that only about five had been studied in sufficient detail for a population viability analysis. Thus, for many species for many years to come, biologists will have to construct mitigation strategies based on less precise tools.

A cruder though more practical means to accomplish the goal of ensuring the environmental integrity of conservation banking could be through use of mitigation ratios that account for the location and importance of various habitats. For example, for habitat that is isolated and contains a small population of the target species, one could foresee a mitigation ratio close to 1:1. On the other hand, the destruction of habitat in areas that are important to recovery, either because of their proximity to existing populations or to areas of public land that currently have or are planned to have large populations, would require larger mitigation ratios (e.g. three acres to be protected at the conservation bank for every acre of habitat taken).

Two aspects of the Mobile conservation bank are relevant to this discussion. First, as noted earlier, in determining the currency used, it was decided that the total number of credits available would correspond to 80 per cent of the carrying capacity of the bank for gopher tortoises. This had the effect of requiring a greater than 1:1 mitigation ratio. That is, for every acre of occupied habitat taken, on average 1.25

acres of habitat would be protected and restored. Second, the bank was specifically designed to compensate for the loss of small, isolated tortoise populations. Indeed, there are few large populations remaining in the county. However, the ESA permit authorizing the bank specifies that it cannot be used to mitigate the loss of larger populations in more pristine habitat. USFWS is also given oversight responsibilities to ensure that this limitation is enforced. In theory, if a Mobile County developer came forward with a proposal to take a large number of tortoises in outstanding longleaf-pine habitat, he would not be able to use the Mobile bank and would, in effect, have to avoid take or find alternative mitigation.

Given the fact that all habitat restoration projects, including conservation banks, involve some measure of uncertainty and unpredictability, one could make the case that all mitigation ratios should substantially exceed 1:1. In other words, developers should always restore more than they destroy, not just to advance the recovery of a given species, but to account for inevitable hitches in restoring habitats and successfully translocating animals. One caution, however, is that as the mitigation ratio rises, the economics of conservation banking become more difficult. In the case of the prairie dog, the cost of producing credits at the bank is relatively inexpensive, so the 2:1 ratio is both scientifically and economically defensible. However, in other instances, if developers or those seeking credits are required to pay enormous sums to conservation banks, then USFWS will be under greater political pressure to help developers skirt the rules or to simply not enforce the ESA. In our experience, this is an important consideration. Mitigation ratios should be set using the best, objective scientific information. (This means, of course, that such ratios should be subject to change over time as new scientific information is collected.) If the best information is used, then issues related to the economics of conservation banking can be openly discussed and evaluated by purchasers, sellers and the public.

Ownership of the bank

The question of who owns the bank might not seem to be relevant to the subject of ecological considerations in conservation banking, but it is. In a number of cases, USFWS has permitted habitat losses on private lands to be mitigated by restoration activities carried out on federal lands. For example, red-cockaded woodpecker mitigation has been allowed on the Croatan, Kisatchie and Ocala national forests. Under section 7 of the ESA, federal lands are already required to be managed for the benefit of endangered species. It is fair to ask, then, whether any real additional habitat protection is gained by allowing mitigation on those lands.

While officials in USFWS contend that conservationists should not care who owns the habitat, in fact, the opposite is true. Mitigation onto federal lands is artificially cheap because the federal government does not have to pay the full cost of habitat protection since the land is already effectively protected from development (no one is building shopping malls on the national forests). In the case of the red-cockaded woodpecker, mitigation costs per breeding pair on federal lands are as low as $1000 to $4000, while on private lands they can run well over $100,000. This creates a perverse incentive in that landowners who might otherwise manage their lands for endangered species can take advantage of cheap mitigation onto federal lands and thereby rid their own property of endangered species. If doing away with ESA regulations is virtually free, why not avail oneself of it?

In the case of the Mobile bank, the land is owned by MAWSS, a public utility. In the case of the UPD, the land is owned by SITLA, a state agency vested with the authority to manage those lands for the benefit of Utah schools. The difference, however, is that neither MAWSS lands nor SITLA lands are federal, and for purposes of the ESA, they are treated exactly the same as private lands. Thus, the landowners (Mobile County and the State of Utah, respectively) bear all the costs of habitat protection and restoration. We see no reason to permit conservation banking onto federal lands.

Timing of credit sales

An important question in the design of conservation banks is when the sale of credits can begin. This is an issue primarily for those conservation banks where habitat restoration is a key activity. Clearly, the best situation from the perspective of the species is one that allows 'take' only after the conservation bank has successfully restored both the habitat and the species in question (with 'success' defined in terms of a reproducing population of the species). That wasn't the case for the red-cockaded woodpeckers that were translocated to federal lands following the loss of their habitat on private lands. Mitigation was deemed successful if the institutions involved in the mitigation effort tried to establish a breeding pair over the course of 3–4 years, whether or not such efforts succeeded.

In wetland banking, it is common for portions of credits to be released based on a performance schedule agreed to with the US Army Corps of Engineers. For example, 30 per cent of credits are released once soil and hydrologic criteria are met, 30 per cent are released once plant populations are established, and the remainder released once colonized by certain indicator species. Because most conservation banks created to date involve the preservation of existing wetlands, the successional release of credits has not been widely used. However, in many cases, bank owners can acquire additional credits after proving that they have substantially improved habitats or demonstrated increased populations of indicator species.

In the case of the UPD bank, once credits for maintaining the existing habitat are sold, SITLA can earn additional credits by increasing the population of UPDs at the bank. But with UPD populations prone to major fluctuations, when should credits become available? The bank addresses this problem by allowing the sale of newly generated credits only after specific population thresholds are passed (e.g. increases of more than 50 adult UPDs during a spring count) for two successive seasons.

In the case of the gopher tortoise, the baseline population at the bank prior to its opening was ten tortoises. Given the long time to maturity and typically low reproductive success of tortoises, growth rates of that population are expected to be very low. So, simply waiting for the population to expand before any credits can be sold is impractical. Also, because of slow growth in tortoise populations, it is important to rescue and translocate tortoises from sites where development will occur. Thus, the bank allows credits to be sold once the habitat is restored and tortoises have been relocated, but it does not require the bankers to wait until the translocated tortoises have reproduced. Some ecologists will probably disagree with this decision, but it does reflect the economic realities of creating a conservation bank for a species like the gopher tortoise. We would argue that the timing of credit sales has to be tailored to the specific needs of the species. Where the biology of the species allows for it, the preference should always be to have the species well-established and reproducing at the bank before credits are sold.

Monitoring

Monitoring must be built into the ongoing operations of each conservation bank – both with respect to compliance with the terms of the agreement and with respect to the effectiveness of all activities related to habitat restoration, management and species restoration. It is the responsibility of USFWS to ensure that conservation banks are carrying out the restoration and management activities required under the permit that each bank receives from the agency. Beyond compliance monitoring, however, it is important that each bank monitors the progress of restoration and management activities. Each conservation bank should also have a habitat restoration and/or population goal that it strives to meet.

In the case of the tortoise, it is instructive to look at how the species has been treated in Florida, where it is not federally protected but where a state law requires that tortoises in the path of development be relocated to safe sites. It appears that relocations in Florida have been haphazard, with little thought given to where tortoises are moved. As a result, many of the translocated tortoises have fared poorly. The poor track record in Florida caused many scientists to view the Mobile bank with scepticism. In response to their concerns, USFWS and MAWSS developed a monitoring programme that includes periodic surveys and radio-telemetry of existing and translocated tortoises at the bank. In addition, the habitat conservation plan establishing the bank empowers a scientific oversight committee to review monitoring results. Monitoring to date has shown that tortoises have, in fact, remained at the site and have begun to reproduce. It bodes well for the future of the tortoise at the bank.

In the case of the UPD, there is uncertainty with respect to the impacts of livestock grazing on prairie-dog populations. Thus, the SITLA bank incorporates a carefully designed study of the effect of different grazing intensities on UPDs. We encourage both bankers and regulators to think of conservation banks as mini-experiments in restoration ecology, and to design the banks and the monitoring programmes with the goal of improving our knowledge about endangered species. Unfortunately, across the country, monitoring of conservation banks is haphazard and limited, and it is not at all clear whether these banks are supplying the habitat for which they were established (J. Fox, pers. comm.).

Permanence

The USFWS policy on conservation banking is that bank habitat must be protected in perpetuity. As a general rule, both management and protection of the habitat at the conservation bank can be made permanent through a conservation easement, deed restriction or similar mechanism. In the case of the UPD, for example, SITLA is required to donate a conservation easement on the three sites within the conservation bank to the Utah Division of Wildlife Resources prior to the sale of credits. In the case of the gopher tortoise, the permit issued by USFWS requires permanent protection, although MAWSS was not required to actually put a deed restriction on the property. Since the property is managed as a buffer for a reservoir, it is not clear that such a restriction is required (but it would certainly not have been harmful).

Are there instances when less-than-permanent protection is adequate? As noted previously, a number of endangered species utilize early-successional habitats created by disturbances such as fire. In the absence of such disturbances, the habitat becomes unsuitable for them. An argument for less-than-permanent protection could be made in cases where permanent protection at the conservation bank raises the costs of mitigation so high that landowners seeking habitat credits are more likely to just wait out the species and allow natural succession to get rid of their ESA liability. If, for example, a conservation bank required 50 years of habitat protection and management for a species that requires prescribed fire every 2–3 years, there might be an argument that recovery would be advanced even in the absence of permanent protection. Such a case might also argue for government assistance in retiring development rights on the property. Thus, mitigation dollars could purchase long-term protection and management while government funding could make the protection permanent. Whether or not such an approach makes sense cannot be decided in the abstract, however; it will depend upon the ecology of the species and its habitat and the economic forces acting on the landowners.

Conclusion

Evident in much of our discussion of ecological issues associated with conservation banking is the inevitable tension between ensuring that banks contribute to the recovery of endangered species and ensuring that they are economically viable. While in theory the prohibitions against habitat destruction under the ESA are severe, USFWS has a spotty record in enforcing the Act. When political pressure on the agency builds, it is more likely to buckle. A well-designed, economically feasible bank can reduce this pressure while simultaneously advancing the conservation of the species in question.

Thus, in designing banks, conservationists should adhere to scientifically defensible guidelines while at the same time striving to ensure that banks are economically viable. Ecological considerations should ultimately revolve around the long-term recovery needs of the species. For those species with up-to-date recovery plans, conservation banking may help to achieve recovery goals. Where recovery plans do not exist or are out-of-date, USFWS should work to design banks in such a way that there is a high degree of certainty that recovery efforts will be bolstered by the activities of the bank. Moreover, all banks should be seen as an opportunity to gain new insights into the management of endangered species and their habitats. Ultimately, everyone – from the regulated community to the environmental community – will benefit if the market for credits is designed in such a way to help move species off the endangered species list.

References

Bauer, M., Fox, J. and Bean, M. (2004) 'Landowners bank on conservation: The U.S. Fish and Wildlife Service's guidance on conservation banking', *Environmental Law Reporter*, vol 34, no 8, pp10717–10722

Bean, M. J. and Dwyer, L. E. (2000) 'Mitigation banking as an endangered species conservation tool', *Environmental Law Reporter*, vol 30, no 7, pp10537–10556

Bonnie, R. F. (1999) 'Endangered species mitigation banking: Promoting recovery through habitat conservation planning under the Endangered Species Act', *The Science of the Total Environment*, vol 240, pp11–19

Bonnie, R. F. (2004) 'From Cone's Folly to Brosnan Forest and beyond: Protecting red-cockaded woodpeckers on private lands', in Costa, R. and Daniels, S. J. (eds) *Red-cockaded Woodpecker: Road to Recovery*, Hancock House Publishers, Blaine, WA, pp163–173

Fox, J. and Nino-Murcia, A. (2005) 'Status of species conservation banking in the U.S.', *Conservation Biology*, vol 19, no 4, pp996–1007

Fox, J., Daily, G. C., Thompson, B., Chan, K. M. A., Davis, A. and Nino-Murcia, A. (2006) 'Conservation banking', in Scott, J. M., Goble, D. D. and Davis, F. W. (eds) *The Endangered Species Act at Thirty: Conserving Biodiversity in Human-dominated Landscapes*, Island Press, Washington DC

Groom, M. J., Meffe, G. K. and Carroll, C. R. (2006) *Principles of Conservation Biology* (3rd edition), Sinauer Associates, Sunderland, MA

Roberts, L. (1993) 'Wetlands trading is a loser's game, say ecologists', *Science*, vol 260 pp1890–1892

Turner, W. R., Wilcove, D. S. and Swain, H. M. (2006) 'Assessing the effectiveness of reserve acquisition programs in protecting rare and threatened species', *Conservation Biology*, vol 20, pp1657–1669

USFWS (1991) 'Utah prairie dog recovery plan', US Fish and Wildlife Service, Denver, CO

Wilcove, D. S. and Chen, L. Y. (1998) 'Habitat management costs for endangered species', *Conservation Biology*, vol 12, pp1405–1407

Wilcove, D. S., Rothstein, D., Dubow, J., Phillips, A. and Losos, E. (1998) 'Quantifying threats to imperiled species in the United States', *BioScience*, vol 48, pp607–615

Zedler, J. (1986) 'Restoring diversity in salt marshes: Can we do it?', in Wilson, E. O. (ed) *Biodiversity*, National Academy Press, Washington DC, pp317–325

6

Legal Considerations

Royal C. Gardner[1]

In the 1978 case of *Tennessee Valley Authority vs Hill*,[2] the US Supreme Court wrestled with the economic impact of the Endangered Species Act (ESA). The TVA's Tellico Dam and Reservoir Project would inundate the lower reaches of the Little Tennessee River and jeopardize the continued existence of the endangered snail darter. Could the presence of this three-inch perch prevent the project's completion, despite the fact that $53 million already invested in the project would be lost? Declaring that the 'plain intent of Congress in enacting this statute was to halt and reverse the trend toward species extinction, *whatever the cost*,'[3] the Court ruled that the tiny fish indeed trumped the federal project. Congress, the Court stated, had concluded that the value of endangered species was 'incalculable'.[4]

After the Court's decision, Congress moderated its view of the value of endangered species. To provide more flexibility, Congress amended the ESA to provide for an exemption that could be granted by a high-level federal committee, informally known as the God Squad. The TVA petitioned for an ESA exemption, which the God Squad denied. The committee chair, Secretary of the Interior Cecil Andrus, remarked, 'I hate to see the snail darter get credit for stopping a project that was ill-conceived and uneconomical in the first place.'[5] Undeterred, project proponents inserted a provision in an appropriations bill waiving ESA requirements for the Tellico Dam, which Congress then passed. The Tellico Dam project went forward, destroying the snail darter's habitat in the Little Tennessee River. In the end, somewhat surprisingly, the economically unjustified project did not extirpate the species, as snail darters were found elsewhere. The snail darter's saga, however, illustrates that despite the lofty rhetoric of Congress and the Supreme Court, there are limits to the costs that society will pay to protect endangered species.

As the ESA was implemented, it became clear that all property owners did not necessarily share a benevolent view of endangered species. Endangered species and their habitat were often seen as liabilities; their presence could impede development and drive down property values. Accordingly, some property owners managed their land to make it less attractive to endangered species or, in extreme cases, adopted the 'shoot, shovel, and shut up' approach.[6] Ironically, the inflexibility of the ESA's protections created a disincentive to manage land for the benefit of endangered species.

Conservation banking attempts to reverse this disincentive by translating an 'incalculable' value into a market value. This chapter will examine the legal issues associated with conservation banking and its objective of creating a system in which endangered species and their habitats are viewed as economically beneficial to property owners.

Legal framework for conservation banking

Like wetland mitigation banking, conservation banking is an odd business. The government controls both the demand for and supply of mitigation credits. Without laws such as federal or state endangered species acts, there would be no market for conservation bank credits.[7] The requirement for developers to provide mitigation to offset their adverse environmental impacts is the regulatory driver that creates the mitigation industry.

The federal ESA and its state counterparts generally follow a similar model, although the terminology may vary. First, protected species must be identified. Under the federal ESA, the US Fish and Wildlife Service (FWS) has responsibility for designating most species, while National Marine Fisheries Service has responsibility for marine and anadromous species.[8] At the federal level, the determination to list a species is supposed to be a purely scientific endeavour, resting 'solely on the basis of the best scientific and commercial data available'.[9] (In this context, 'commercial' data refers to the over-utilization of a species for commercial purposes.) An agency can also designate critical habitat for the species, an area deemed to be necessary for conservation of the listed species.[10] Here, however, the federal decision to designate critical habitat must consider the economic consequences of the designation. The federal ESA spells out the process by which an agency can list a species and designate critical habitat: notice-and-comment rulemaking, which results in a regulation.

Once a species is listed as endangered or threatened (or, in some states, the terminology is rare, unusual or of special concern), it receives certain protections. For example, the federal ESA prohibits the 'taking' of any listed species.[11] 'Take' is defined broadly in the statute as 'harass, harm, pursue, hunt, shoot, wound, kill, trap, capture, or collect, or to attempt to engage in any such conduct'.[12] The FWS – through a regulation after public notice-and-comment rulemaking – has further defined 'harm' to include significant habitat modification or degradation that results in actual injury or death to wildlife.[13] The US Supreme Court upheld the validity of this regulation in *Babbitt vs Sweet Home Chapter of Communities for a Great Oregon*.[14] Accordingly, if someone 'takes' a listed species, he or she (or it in the case of a corporation) may be subject to civil and/or criminal penalties.

The prohibitions against 'taking' a species are not absolute, however. If a federal agency's proposed activities may affect a listed species, the agency is obligated to consult with the FWS under ESA section 7(a)(2). The point of the inter-agency discussion is to ensure that the proposed activity will not jeopardize the species' continued existence or adversely modify its critical habitat. If the FWS finds that the proposed activity will not jeopardize the species' continued existence, it will issue a 'no jeopardy' opinion. Yet the agency's proposed action could nevertheless result in a 'take' of the protected species, perhaps resulting in the deaths of several individual animals. In such cases, the FWS may issue an incidental take statement to allow the activity to proceed. The incidental take statement may include mitigation conditions. So long as the agency and its employees comply with the terms of the incidental take statement, they will not be subject to ESA sanctions.

A slightly different process is available for non-federal actions that may affect a listed species or its habitat. ESA section 10(a) authorizes the FWS to grant a permit for activities that would result in an incidental take of a listed species. The permittee must, 'to the maximum extent practicable, minimize and mitigate the impacts of such taking'[15] and thus must submit a Habitat Conservation Plan that includes mitigation provisions. The section 10 permit will contain mitigation conditions that the FWS imposes. Many states have similar requirements. In the end, the result is the same for public and private actors: there is a need for a project proponent to provide mitigation for adverse impacts to protected species.

Increasingly, credits from conservation banks are being used to provide this required mitigation. However, a search of the text of the federal ESA for any reference to 'conservation banking' will only yield disappointment. Congress has not enacted a law expressly authorizing (or, for that matter, forbidding) conservation banking. A review of the FWS regulations is also fruitless; there are no federal regulations explicitly discussing conservation banking. It is only when one digs down to the level of guidance documents will references to conservation banking be unearthed.[16]

Legal distinctions between statutes, regulations and guidance (policy)

To understand the legal framework of conservation banking, one must appreciate the difference between statutes, regulations and guidance (also known as policy). A statute is a written enactment of Congress or a state legislature, typically approved by the President or a governor. When someone refers to the 'law', he or she is probably referring to a statute, such as the ESA.

Yet a regulation, properly promulgated, also has the force of law.[17] Its prohibitions and proscriptions are binding on agencies and the public just as a duly enacted statute is. A regulation, however, is not drafted by elected lawmakers; rather, regulations are generally written by unelected government employees (bureaucrats). Although these government officials may have the scientific and technical background necessary for crafting an effective regulation, they lack the political legitimacy to do so by themselves. Yet Congress (and state legislatures) cannot anticipate all the details with respect to the implementation of a particular law, and it therefore delegates the responsibility to fill in the gaps to agency officials.

To resolve the concern about unelected bureaucrats issuing regulations with the force of law, there are several checks on an agency's discretion. First, the career government employees (who typically have the technical expertise) are accountable to political appointees. For example, the Director of the US Fish and Wildlife Service is appointed by the President, and the Director of the California Fish and Game Department is appointed by the governor. A proposed regulation will be reviewed by these appointees (or lower-level appointees), providing at least an indirect link between an elected official and the drafter of a regulation. At the federal level, regulations must also be reviewed by the Office of Management and Budget, in part to ensure that the rules are consistent with the administration's policies. The second way a regulation is infused with political legitimacy is through the method by which it is promulgated.

Regulations that have the force of law almost always must go through a public notice and comment process. Under the federal ESA, when the FWS wishes to list a species as threatened or endangered, it must issue a notice in the Federal Register along with the text of the proposed regulation.[18] State and local agencies, scientific organizations and the general public are invited to provide comments. The agency will hold a public hearing on the proposed regulation if requested. The agency must then consider the comments that it has received (consistent with the statutory criteria, of course). After evaluating this input, the agency may then proceed to issue a final rule or regulation, which is also published in the Federal Register. The public comment process helps the agency in developing more effective rules, and it also serves to legitimize the role of unelected officials promulgating binding regulations.

Promulgating a regulation through the notice-and-comment process can be cumbersome, and agencies will often opt to issue guidance instead. Guidance can be issued from most levels within an agency (e.g. from headquarters or a field office),[19] and typically the guidance can be issued with little or no public involvement. The guidance may have various labels: a policy document, memorandum of agreement, regulatory guidance letter, interpretive rule, etc. Whatever the title, these documents share a common trait: they do not have the force of law. They may inform the public and the regulated community about how an agency will interpret a statute and its regulations, but they are not *law* in the sense of a statute or regulation. As such, agency field personnel may not always feel compelled to follow the guidance.[20]

There is nothing intrinsically wrong or improper with such guidance documents. Indeed, federal agencies are authorized to issue such policy statements and interpretive rules under the Administrative Procedure Act.[21] But because they are not necessarily subjected to a public notice-and-comment process and because they are not codified in the Code of Federal Regulations, these agency pronouncements are not binding regulations. They typically do not create any enforceable rights or obligations. A court will not grant as much deference to guidance documents as it would to a regulation promulgated through notice and comment. Moreover, because guidance documents can be issued without public notice and comment, they can be modified or revoked without public notice and comment.

At the state level, the authorization for conservation banking also is primarily found in guidance documents. While state statutory and regulatory authorizations for wetland mitigation banking are plentiful,[22] explicit references to conservation banking are few. Hawaii's statute on 'habitat banking', which allows a project proponent to mitigate endangered species impacts by the purchase and preservation of property on which the affected species are found, would seem to permit privately operated conservation banks.[23] Yet in California, the state with the most active conservation banking programme, the rules governing conservation banking are set forth as guidance in a policy document.[24]

Thus, the conservation banking industry is primarily built on guidance, not statute or regulation. Yet the lack of a firm legal foundation does not necessarily preclude banking; an industry can flourish under a guidance document that specifies how banks are to be established and operated, as the experience of federally approved wetland mitigation banks demonstrates.[25] Nevertheless, the legal status of conservation banking at the federal and state levels is an element of risk that must be taken into account by those wishing to enter the market. The risk is reduced if conservation banking is expressly authorized by regulation or statute, as any changes to a regulation or statute will allow for more public participation in the process.

An introduction to conservation easements

In many first-year property law classes, students are told that the ownership of land can be viewed as a 'bundle of sticks'. Each stick in the bundle corresponds with a particular right, such as 'the right to occupy the land (for the most part), to exclude others (nearly always), and to decide whether to build on the land'.[26] A fee owner (or an owner in fee simple) possesses the maximum rights in a particular parcel of land. In the absence of regulatory constraints, the owner could use the property however it wished.

Another right in the bundle is the right to alienate or convey the land. An owner could sell the parcel in its entirety, effectively transferring all the sticks in the bundle. The owner may instead choose to physically subdivide the parcel and sell a portion of the property. The owner could also decide to transfer particular rights in the parcel, such as conveying the right to extract subsurface minerals, while retaining surface rights. The variations are limitless.

The primary objective of conservation banking is to offset impacts to endangered species habitat in perpetuity. Accordingly, who owns the land and who otherwise possesses rights in the land are critical issues. Assume that a conservation bank site that provides habitat for Florida scrub jay is owned in fee simple by a person or entity that is committed to the long-term protection of the site. Good intentions, however, are not legally binding, and circumstances can change one's position. What if the person dies and the heirs wish to sell the property to a developer? What if the entity runs into financial difficulties and likewise wishes to sell the property to a developer? Ordinarily, a fee owner may do so. Obviously, such a transaction would defeat the objective of conservation banking, and thus some legal encumbrance should be placed on the property to prevent this scenario. A conservation banking agreement (CBA) will typically require that a conservation easement (or servitude) be placed on the bank site to protect the property's conservation values for the long term.

A conservation easement is defined as 'a legal agreement between a landowner and an eligible organization that restricts future activities on the land to protect its conservation values'.[27] In essence, the owner has conveyed all or part of one or two sticks in the bundle of rights: the right to use the property (by promising to forgo certain activities) and the right to exclusive possession (by permitting an easement holder to monitor and inspect the site).

Conservation easements are a relatively recent development in property law. Traditionally, easements were limited to the owner of an adjacent property and involved 'rights of access and protection of views'.[28] State legislatures expanded upon the traditional concept of easements to permit 'conservation easements to be created on land without an adjacent "benefited parcel" and to last forever'.[29] Accordingly, conservation easements are creatures of state law – specifically state statutes. Because statutes authorizing conservation easements can vary from state to state, it is important to review the rules in each individual jurisdiction.[30]

The terms of the conservation easement must be drafted with care and precision. While the immediate parties may have no disagreement about the permitted and prohibited activities (e.g. is hunting and selective timbering permitted, and if so to what extent?), the conservation easement should last beyond the lives of the individuals involved. Forever, as they say, is a very long time. Accordingly, the language of the conservation easement for a conservation bank site should be written with other parties in mind. Are the provisions clear enough to be understood and enforced by future landowners and easement holders?

The holder of a conservation easement has a right to limit activities on the easement site to protect its conservation values. Yet one who holds a right may decline to exercise that right. In a different context, persons accused of a crime routinely waive their right against self-incrimination or their right to an attorney. Ordinarily, an easement holder may similarly waive its rights and decline to enforce the terms of the easement. One aspect of having a right is having the discretion about whether to assert it.

The holder of a conservation easement for a conservation bank site, however, may have more than a mere *right*. The easement holder also has a *duty* to protect the site's conservation values, especially when the grantor of the easement has realized federal tax benefits.[31] If the holder of a conservation easement fails in its duty to protect the site (e.g. passively allows development of the site or trespassers to damage the site), then the conservation organization may not be eligible to receive such tax-deductible easements in the future.[32]

In contrast to traditional easements, the terms of a conservation easement (in some states) may be enforced by someone other than the easement holder. Because a conservation easement provides public benefits, some states authorize their attorney general to bring enforcement actions in court or to intervene in easement enforcement litigation.[33] Furthermore, some conservation easements specify a backup holder, who is 'empowered to take over an easement if the original holder can no longer manage it'.[34] A backup holder also can be granted a right of enforcement. Still, third parties and members of the general public do not have legal standing to enforce the terms of a conservation easement.[35]

Are conservation easements forever?

The law generally discourages permanent limitations on the use of land.[36] Although conservation easements are intended to last in perpetuity, there are a number of threats that may frustrate this goal. The threats may arise through operation of law (recording laws), actions by government (eminent domain), exercise of superior rights (such as a pre-existing mortgage) or a change in conditions (the site no longer provides its intended conservation value). All of these threats may frustrate the protection of a conservation bank site.

Like most easements, a conservation easement should be recorded with the appropriate registry of deeds. The easement's recordation provides notice to subsequent purchasers of the property. If the easement is not properly recorded, the holder may not be able to enforce it against a subsequent landowner. Many states also have a 'marketable title act', legislation that requires an easement holder to periodically re-record the conservation easement.[37] If an inattentive conservation easement holder fails to re-record within the prescribed period, the easement may be void. Whether a conservation easement for a conservation bank site must be periodically re-recorded depends on a particular state's law.

An easement is a property right, and as such, it can be subject to eminent domain. Under the Fifth Amendment to the US Constitution (which applies to both the federal and state governments), the government may take one's property for a public purpose, so long as it pays just compensation to the owner. Thus, the government may condemn and acquire private property for a highway project, although its eminent domain powers are much broader than that, as the US Supreme Court noted in 2005 in *Kelo vs City of New London*.[38] When the government acquires property by eminent domain, it may extinguish any conservation easements. This recently transpired in Florida in the context of a wetland mitigation bank, where the state planned to build a parkway through a portion of the Wekiva River Mitigation Bank. While the state and the mitigation banker reached an agreement regarding compensation, the events underscore the notion that the government can choose to acquire – and invalidate – a conservation easement.[39]

A conservation easement may also be terminated through a mortgage foreclosure. A mortgage on a property creates a lien, and a subsequent conservation easement does not trump. The sale of the property at a foreclosure sale would extinguish the conservation easement. Accordingly, when obtaining the easement, the easement holder should secure a 'mortgage subordination' from the lender to ensure that the conservation easement would continue in the event of a foreclosure.[40] Internal Revenue Service regulations require subordination of mortgage liens when the property owner is seeking federal tax benefits.[41]

Changing land conditions can threaten the viability of species, but they can also call into question the vitality of restrictions on real property. The doctrine of changed circumstances is sometimes invoked to modify long-term arrangements, including easements.[42] A landowner may argue that the surrounding area has changed so much that it makes no sense to continue to burden the property with the easement. While this doctrine should not generally apply to conservation easements,[43] one can envision its applicability to a conservation easement that is narrowly written. What if the sole purpose of a conservation easement is to protect habitat to ensure the conservation of the Dehli Sands Flower-Loving fly or the Florida panther, and one of these species is later declared extinct? If the only purpose of the easement is to assist with the survival of a species, and that species no longer exists, then the landowner may convince a court to terminate the conservation easement. Such a possibility counsels that conservation easements be drafted for multiple purposes.

Enforcement of conservation banking agreements

In one sense, a conservation banker sells credits, which may be quantified in 'acres of suitable habitat, or numbers of individuals or breeding pairs of the species supported by the bank'.[44] But what the conservation banker is really selling is the release of legal liability under the ESA; the permittee as purchaser of the bank credits is no longer responsible for the success of the mitigation.[45] With the sale of the credits, the conservation banker is now responsible for the mitigation in accordance with the terms of the CBA. If the conservation banker violates the agreement, is it legally liable? The answer is not entirely clear, in part because of the legal framework of conservation banking.

Unlike a permittee's responsibility for its HCP, a conservation banker's responsibility is not specifically mentioned in the statutory language of the ESA. Nor is the conservation banker's responsibility spelled out in federal regulations. It is only in guidance documents where one finds a discussion about the conservation banker's responsibility and liability.[46] Yet an agency may not commence an enforcement action based solely on the violation of guidance. It must point to a violation of statute or regulation.

This is not to say that the government lacks legal remedies if a conservation banker violates its agreement. Although an agency may have difficulty proceeding with an enforcement action under the ESA, it could bring a breach of contract, unjust enrichment or similar action against the conservation banker. A recent case involving a failed wetland mitigation bank in Kentucky illustrates this point. When the wetland mitigation bank did not fulfil its obligations under an agreement with federal and state agencies, the federal government did not bring a Clean Water Act enforcement action, presumably because the bank was not a permittee. Instead, the government alleged that the bank was liable for breach of contract, negligent misrepresentation, breach of duty of good faith and fair dealing, and unjust enrichment.[47] The corporate and individual defendants settled the case by agreeing to pay $70,000 to the Kentucky Department of Fish and Wildlife Resources.[48]

Conservation banks and bankruptcy

Assume that a conservation bank has been established for the benefit of the Florida scrub jay. A conservation organization holds the conservation easement and has the responsibility to conduct prescribed burns on the site. It also has the responsibility to prevent trespassers from using all-terrain vehicles on the site. What if the organization has the will to fulfil its responsibilities, but lacks the financial resources to do so? In the worst-case scenario, a conservation organization's financial condition may be so dire that it files for bankruptcy protection, which obviously can have significant consequences for the long-term stewardship of any conservation bank site for which it has obligations.[49]

An entity filing for bankruptcy in federal court may choose to proceed under chapter 7 or chapter 11 of the Bankruptcy Code. Chapter 11 provides the debtor an opportunity to shed certain obligations (including debts), reorganize and make a fresh start. An action under chapter 7 results in the liquidation of the business. Under chapter 11, the debtor will typically submit a plan of reorganization for approval by the bankruptcy judge. Under chapter 7, a government-appointed trustee will wind up the entity's affairs (including distributing proceeds) in accordance with the Bankruptcy Code's provisions. In either case, creditors will seek to be paid; often, however, unsecured creditors receive little or nothing.

Bankruptcy is a powerful instrument, and bankruptcy judges wield tremendous power. In an era where bankruptcy courts will void the pension guarantees promised to a company's retirees or modify the terms of a collective bargaining agreement, it is not surprising to find that a bankruptcy action can also affect environmental rights and obligations, including conservation easements. The bankruptcy court has the power to invalidate the conservation easement and eliminate an entity's long-term stewardship obligations.

There have been several bankruptcy actions that have involved wetland mitigation banks and conservation banks. In one case, the State of New Jersey brought an administrative enforcement action against a mitigation banker (that was in bankruptcy) after it inadvertently drained 19 acres of wetlands.[50] New Jersey contended that its efforts to seek restoration and a fine were the acts of a sovereign, and thus not subject to the bankruptcy court's jurisdiction. The court saw the matter differently, finding New Jersey to be an unsecured creditor and ordering the dismissal of the enforcement action.[51] Another wetland case in Florida had a happier result when the debtor and creditors reached a settlement, which seem to secure the long-term prospects of the mitigation sites.[52] Both of these cases involved for-profit companies filing for bankruptcy protection; yet non-profit entities can go belly up as well, as demonstrated by The Environmental Trust (TET).

TET, a non-profit public benefit corporation, was created in 1990. Fifteen years later, TET was responsible for the management of more than 4600 acres in California, including ten conservation banks in San Diego County.[53] Unfortunately, due to chronic underfunding of its long-term stewardship obligations and the commingling of its endowment funds with operational funds (among other reasons), TET could not fulfil its environmental responsibilities and fired its non-management personnel in 2004.[54] It then filed for chapter 7 bankruptcy, which is not a mere reorganization, but a liquidation of the entire business.

The court-approved liquidation plan contemplates offering TET's properties, along with its maintenance, monitoring and management obligations, serially to interested parties. First, permittees are given the opportunity to assume these obligations (a result that seems unlikely as they paid TET precisely to be released from such burdens). Next, the city and county in which the site is located will be given the option to assume these responsibilities. If the local governments decline, then it is on to the FWS, California Department of Fish and Game, non-profit entities and, finally, the State of California. If all decline, the bankruptcy plan states that TET's interests 'will be abandoned'.[55]

This is not a happy thought, but abandonment of these sites is not a foregone conclusion. The matter is still in court. Nevertheless, TET provides a cautionary tale about long-term stewardship responsibilities.

There are several ways to avoid this scenario. First, as discussed in greater detail in Chapter 9, the endowment accounts must be adequately funded for perpetual maintenance, monitoring and management. This is the long-term steward's responsibility in the first instance, but the regulatory agencies also must exercise rigorous oversight. Non-profit corporations should not receive a pass simply because of their status. Second, the conservation easement should specify a co-holder or a backup holder in case the entity with primary responsibility encounters difficulties. The terms of the easement would dictate who would bear the continuing obligations of protecting the site's conservation values, rather than leaving the matter to the bankruptcy court.

The other 'take': conservation banks and regulatory takings

From an environmental perspective, the benefit of conservation banking is the establishment of larger reserves and the greater connectivity of habitat. Conservation banking may benefit permit applicants by saving them time and money. Furthermore, for owners of endangered species habitat, conservation banking provides 'an opportunity to generate income from what may have previously been considered a liability'.[56] This last benefit also offers an ancillary benefit to government agencies: it can insulate them from regulatory takings claims.

When the government seizes or occupies one's land, it must pay the landowner just compensation for the physical taking. Even slight physical invasions, if permanent, trigger the just compensation requirement.[57] In addition, it has long been recognized that sometimes government regulation goes 'too far' and constitutes a regulatory taking.[58] In a regulatory taking, the government has not physically seized or occupied the property, but it has interfered with a property owner's rights to such an extent that it is treated as the equivalent of a physical taking.[59] The denial of wetland and endangered species permits sometimes gives rise to a regulatory takings claim. Indeed, private property rights groups focus much of their animus on wetland and endangered species laws.

There are several approaches that a court may take to consider whether a permit denial constitutes a regulatory taking. Typically, a court will conduct an ad hoc, factual inquiry, and the decision will turn on the particular circumstances of each case. In *Penn Central Transportation Co. vs New York*, the US Supreme Court articulated several factors that are relevant to such an inquiry: the economic impact of the permit denial on the property owner; the extent to which the permit denial interfered with reasonable, investment-backed expectations; and the character of the government's action.[60] Under the *Penn Central* test, no one factor is necessarily dispositive.

Although government agencies occasionally lose a regulatory takings case, it is difficult for a property owner to prevail under the *Penn Central* test.[61] The government will point out that there are alternative uses to the property, that limited development can be permitted and the property still retains some economic value. If, however, the property owner can establish that the government's action 'denies all economically beneficial or productive use of land', then the *Penn Central* balancing test will not apply. Instead, the court considers the permit denial a categorical taking or a taking *per se*. The property owner is in a much stronger position under this approach, which the US Supreme Court enunciated in *Lucas vs South Carolina Coastal Council*.[62]

In *Lucas*, a developer purchased two beachfront lots for approximately $1 million. Two years later South Carolina enacted a law that prohibited the construction of habitable improvements within a certain erosion zone. The two lots fell within that zone, and the developer could do nothing with the property. He then filed a takings claim in the South Carolina Court of Common Pleas. The trial court found that the lots were 'valueless' and ordered the state to pay him just compensation.

When the case reached the US Supreme Court, the Court noted that a takings claim usually involves an ad hoc, factual inquiry. The Court emphasized, however, that such an approach is not necessary when government conduct has rendered a property valueless. In such a case, 'when the owner of real property has been called upon to sacrifice *all* economically beneficial uses in the name of the common good, that is, to leave his property economically idle, he has suffered a taking'.[63] After drawing this bright line, the Court carved out a narrow exception. The government would not have to pay just compensation if it was merely acting to prevent a nuisance.

Lucas represented a potentially significant threat to the ESA. If a property owner was instructed by the FWS to leave his property idle for the benefit of a listed species, it could be a taking *per se*. Furthermore, the protection of such species would not be considered to be preventing a nuisance. The government would have to compensate the owner. If there were many such cases, the government might decide that the ESA was too expensive to enforce.

The advent of conservation banking renders this concern academic. If an agency denies a property owner a permit to develop property because of the presence of endangered species, the property is not 'valueless' nor is it 'economically idle'. Rather, the property may have an economic use and value as a conservation bank site. Accordingly, the *Lucas* test would not be applicable to an endangered species case.[64] Instead, a court would turn to the *Penn Central* factors, which are more in the agency's favour. The purpose of conservation banking may be to protect endangered and threatened species, but it also helps protect federal and state endangered species laws themselves.

Conclusion

In the end, conservation banking largely exists in the shadow of the law (i.e. in guidance and policy documents), rather than in statutes and regulations. As such, the rules governing conservation banking are more susceptible to political change. But as wetland mitigation banking has demonstrated, an entire industry can arise relying on agency guidance. Moreover, the fact that the rules are guidance or policy means that the penalties under the ESA will not be triggered if the conservation banker does not abide by the conservation agreement. Instead, the government will probably rely on other enforcement options, such as a breach of contract or unjust enrichment claim.

From a regulator's perspective, it is critical to examine carefully the state laws governing conservation easements. Although conservation easements for conservation bank sites are intended to last in perpetuity, there are a number of events that could frustrate this objective. Just as important, the regulatory agency must ensure that the conservation banker is financially sound. A conservation banker that files for bankruptcy can threaten the long-term conservation of the site. Finally, an unintended benefit of conservation banking is that it can help insulate agencies from regulatory takings claims by creating an economic value in endangered species habitat.

Notes

1 Professor of Law and Director, Institute for Biodiversity Law and Policy, Stetson University College of Law, Gulfport, Florida. The author thanks Kristine P. Jones, a May 2007 graduate of Stetson University College of Law, for her excellent research assistance.
2 437 US 153 (1978).
3 Ibid at 184 (emphasis added).
4 Ibid at 187–188.
5 Mann and Plummer (1995).
6 Reiland (2004) (discussing property owners' reactions to ESA-imposed land-use restrictions). See also Ruhl (2004) (observing that the ESA seems to be a 'perverse statute' that sends the 'wrong message to landowners about what it means to have habitat for endangered species on their property: get rid of it before the government knows it's there').
7 The experience of the wetland mitigation banking industry is illustrative. After a US Supreme Court decision in 2001 that restricted the federal government's ability to regulate impacts to isolated wetlands and other waters, sales volume and gross income of wetland mitigation credits in the Chicago area were cut approximately in half. See Robertson (2006) (noting 'a gross income decline of 46.1% among Chicago bankers between 2000 and 2001').
8 For convenience sake, this chapter will refer only to FWS when discussing the federal ESA.
9 ESA §4(b)(1)(A), 16 U.S.C. §1533(b)(1)(A) (2007).
10 ESA §4(b)(2), 16 U.S.C. §1533(b)(2).
11 ESA §9(a), 16 U.S.C. §1538(a).
12 ESA §3(19), 16 U.S.C. §1532(19).
13 50 CFR §17.3 (2007). Actual injury may result from acts that significantly impair 'essential behavioral patterns, including breeding, feeding, or sheltering'. Ibid.
14 515 U.S. 687 (1995).
15 ESA §10(a)(2)(B)(ii), 16 U.S.C. §1539(a)(2)(B)(ii).
16 See US Fish and Wildlife Service (2003) [hereinafter Federal Conservation Banking Guidance].

17 For a discussion of federal rulemaking and the difference between legislative rules (regulations) and other rules, see Gardner (1991).

18 ESA §4(b)(3-6), 16 U.S.C. §1533(b)(3-6). The FWS also has limited authority to issue an emergency listing without prior public notice and comment: ESA §4(b)(7), 16 U.S.C. §1533(b)(7).

19 For example, in the wetland mitigation context, USACE has issued guidance from the headquarters level, but many Corps districts also promulgate mitigation policies that take into account local circumstances.

20 The pirate Barbossa's refusal to take Elizabeth to shore in the first instalment of *Pirates of the Caribbean* illustrates this point:

> *First, your return to shore was not part of our negotiations nor our agreement so I must do nothing. And secondly, you must be a pirate for the pirate's code to apply and you're not.* **And thirdly, the code is more what you'd call 'guidelines' than actual rules.** *Welcome aboard the Black Pearl, Miss Turner.*

Memorable Quotes for Pirates of the Caribbean: The Curse of the Black Pearl (2003), http://www.imdb.com/title/tt0325980/quotes (emphasis added). See also *National Mitigation Banking Association vs US Army Corps of Engineers*, 2007 (noting that internal guidance is not a binding regulation in a lawsuit alleging that the Corps failed to follow its in-lieu fee guidance for the expansion of the O'Hare Airport).

21 5 U.S.C. §553(b)(A) (2007).

22 Envtl. Law Inst. (2002) (listing 23 states with wetland mitigation banking laws and/or regulations).

23 See Hawaii Revised Statutes (2006). Some states have authorized their transportation agencies to participate in conservation banks. See, for example, Arizona Revised Statutes (2007) (Arizona law permitting the purchase of 'banking credits' as environmental mitigation to offset impacts of transportation facilities); Revised Code of Washington Annotated (2007) (Washington law permitting the purchase and selling of 'mitigation bank credits' as 'environmental mitigation of transportation projects').

24 See California Resources Agency and California Environmental Protection Agency (1995); US Fish and Wildlife Service and California Department of Fish and Game (1996); and California Department of Fish and Game (2002). California's regulations do make a passing reference to conservation banking in the context of water quality certifications, but they do not spell out in any detail the rules governing the establishment, use and operation of conservation banks. See California Administrative Code 23 (2007) (stating that a completed water quality certification must include a compensatory mitigation plan, which can include purchases from a conservation bank).

25 From 1992 to 2005, the number of wetland mitigation banks increased from 46 to 405, despite the fact that the industry was governed through a 1995 guidance document. See Wilkinson and Thompson (2006). In 1998, however, Congress did ratify the guidance by establishing a preference for approved

mitigation banks to be used to offset wetland impacts associated with federally funded transportation projects (Transportation Equity Act for the 21st century, US Public Law (1998), No 105–178, 112 Stat. 107). In March 2006, USACE and US Environmental Protection Agency issued a proposed regulation in the Federal Register on compensatory mitigation for impacts to aquatic resources. The regulation, if and when it is finalized, will trump the guidance.

26 Boudreaux (2005).

27 Byers and Ponte (2005).

28 See note 10.

29 See note 11. Although some states enacted conservation easement statutes in the 1950s, most did so after the creation of the Uniform Conservation Easement Act, a model law proposed by the Uniform Commission on State Laws. See Mahoney (2002).

30 It is also important, prior to acquiring any interest in property, including a conservation easement, to conduct a reasonable inspection to determine whether the site contains any hazardous material. A federal statute, the Comprehensive Environmental Response, Compensation and Liability Act (CERCLA), imposes strict liability on the owners and operators of such sites for the remediation of hazardous material: CERCLA §107(a), 42 U.S.C. §9607(a) (2007). Strict liability means liability without regard to fault; one need not have done anything wrong to be held financially responsible. Furthermore, CERCLA sites are not necessarily located only in industrial areas. There are numerous cases where ranches and farms have been the subject of CERCLA liability, the types of properties that are frequently considered as conservation bank sites. For example, *Western Properties Serv. Corp. vs Shell Oil Co.*, 358 F.3d 678 (9th Cir. 2004) (rancher permitted dumping of 'acid tar'); *Castlerock Estates, Inc. vs Estate of Markham*, 871 F. Supp. 360 (N.D. Cal. 1994) (ranch's cattle-dipping operation caused contamination); *O'Neil vs Picillo*, 883 F.2d 176 (1st Cir. 1989) (pig farm used as a toxic disposal site).

While merely holding a conservation easement should not trigger CERCLA liability, the terms of the conservation easement may provide for the easement holder to manage the property, in which case the easement holder could be considered an operator of the site: Byers and Ponte (2005), p59; Draper (2004) (noting that the 'more active role undertaken by the conservation holder … the more likely the holder can be considered an "operator"'). An easement holder of a contaminated site who had conducted a reasonable inspection of the site may be able to invoke the 'innocent landowner' defence to avoid liability, but the expenses associated with litigation can be significant, even crippling. Thus, a potential easement holder must take these risks into account if an environmental assessment or investigation suggests the presence of hazardous material.

31 A landowner may decide to part with a stick from the bundle of rights – agreeing not to use the property in a certain manner – for a variety of reasons, including financial. The easement holder may pay the landowner fair market value for the easement. If the conservation easement is donated in whole or in part, the landowner may realize a tax benefit. For example, Internal Revenue

Service regulations (which went through a public notice-and-comment process) permit a landowner to deduct the value of the easement from federal income taxes when donated or sold below market value to a qualified governmental unit or non-profit conservation organization. See 26 CFR §1.170A-14 (2007). A landowner who grants an easement may also realize state, local and estate tax benefits. See generally Byers and Ponte (2005), pp93–99.

32 See Byers and Ponte (2005), p157.

33 For example, C.G.S.A. §47-42c (2007) (Connecticut statute permitting the Attorney General to 'bring an action in the Superior Court to enforce the public interest in such [conservation and preservation] restrictions').

34 Byers and Ponte (2005), p170.

35 See Brown (2005).

36 See Klass (2006) (discussing common law barriers to conservation easements and the 'historic mistrust of allowing living owners to exert "dead hand" control over future use of land').

37 Byers and Ponte (2005), p193.

38 545 US 469 (2005) (holding that exercising eminent domain to transfer private land to another private party under an economic redevelopment plan is constitutional).

39 See Florida Department of Environmental Protection (2005). Foreclosure for tax liens presents a similar challenge to the continuing validity of a conservation easement. See Byers and Ponte (2005), p193.

40 See Byers and Ponte (2005), pp60–61.

41 See 26 CFR §1.170A-14(g)(2) (2007). An entity acquiring a conservation easement should also examine whether there are other pre-existing rights or liens, such as mineral rights, that could affect the conservation values of the site. See Byers and Ponte (2005), p88 (discussing tax implications of reservation of mineral rights).

42 See American Law Institute (2000).

43 Id §7.11 (2000).

44 Fox et al (2006).

45 If a permittee violates the terms of its ESA permit (including mitigation conditions or the HCP), the ESA authorizes the assessment of civil and/or criminal penalties. ESA §11, 16 U.S.C. §1540.

46 Significantly, the federal conservation banking guidance does not suggest that violations of CBAs will lead to an enforcement action under the ESA. Instead, each agreement must contain 'provisions for a dispute resolution process'. Federal Conservation Banking Guidance, *supra* note 16, at p14.

47 See *United States vs Hawkins* (2005).

48 See *United States vs Hawkins* (2006).

49 For a detailed discussion on this topic, see Gardner and Radwan (2005).

50 Id, pp10597–10599.

51 Id, p10599.

52 Gardner and Radwan (2006) (updating earlier article). While the future of the banks in Florida seems secure, prospects are much more uncertain for a

wetland mitigation bank in North Carolina, which was also owned by the same company. See ibid.

53 California Environmental Resources Evaluation Centre (CERES) identifies TET as the conservation banker operator/owner and long-term operator of the following: Boden Canyon; Crestridge Habitat Management Area; Lake Hodges; McGinty Mountain; Marron Valley Preserve; O'Neal Canyon; Poway (SANREX) Mitigation Land Bank; and Upham Vernal Pools (CERES, 2007). TET is listed as the operator/owner of two other sites – Ramona Vernal Pool Preserve and San Vicente Conservation Bank – but does not have long-term stewardship responsibilities for them (Ibid). The ten sites have a combined area of 3337 acres, and credits were based on a one-acre-to-one-credit ratio. CERES reports that approximately 750 credits had been utilized. (See ibid.)

54 Combined Disclosure Statement and Liquidating Plan of Reorganization, dated as of 20 December 2005, pp9–10, *In re The Environmental Trust, Inc.*, No. 05-02321-LA11 (Bankr. S.D. Cal.).

55 Ibid, p6.

56 Federal Conservation Banking Guidance, *supra* note 16, p1.

57 For example, *Loretto vs Teleprompter Manhattan CATV Corp.* (1982) (holding that government-authorized installation of cables constituted a physical taking).

58 *Pennsylvania Coal Co. vs Mahon* (1922).

59 See *Lucas vs South Carolina Coastal Council* (1992).

60 *Penn Central Transp. Co. vs New York* (1978).

61 Even a property rights 'victory' may not result in the owner receiving compensation. Consider, for example, the case of *Palazzolo vs Rhode Island.* Mr Palazzolo sought just compensation because he was not permitted to fill in property that contained wetlands. He lost in Rhode Island courts and appealed to the US Supreme Court. Although Palazzolo won a reversal in the Supreme Court, 533 US 606 (2001), he still lost on remand. The trial court again rejected his takings claim, reasoning that 'Constitutional law does not require the state to guarantee a bad investment.'

62 505 US 1003, 1017 (1992).

63 Ibid at 1019.

64 This argument was also advanced in the context of wetland mitigation banking. See Gardner (1996).

References

American Law Institute (2000) *Restatement (Third) of Prop.: Servitudes*, American Law Institute Publishers, St Paul, MN

Arizona Revised Statutes (2007) 28:7092. Land acquisition; transportation purposes

Babbitt vs Sweet Home Chapter of Communities for a Great Oregon (1995) 515 US 687, US Supreme Court

Boudreaux, P. (2005) 'The three levels of ownership: Rethinking our restrictive homebuilding laws', *Urban Lawyer*, vol 37, pp385–402

Brown, C. N. (2005) 'A time to preserve: A call for formal private-party rights in perpetual conservation easements', *Georgia Law Review*, vol 40, pp85–152

Byers, E. and Ponte, K. M. (2005) *The Conservation Easement Handbook*, Land Trust Alliance, Washington DC

California Administrative Code (2007) 23:3856(h)(5). Water quality certification

California Department of Fish and Game (2002) *Habitat Conservation Planning Branch, Lands Not Appropriate for Conservation/Mitigation Banking*, www.dfg.ca.gov/hcpb/conplan/mitbank/mitbank_policies/cmb_notaccept.shtml, accessed April 2007

California Environmental Resources Evaluation System (2007) *Conservation Banks in San Diego County*, http://ceres.ca.gov/topic/banking/san_diego.html, accessed April 2007

California Resources Agency and California Environmental Protection Agency (1995) *Official Policy on Conservation Banks*, http://ceres.ca.gov/wetlands/policies/mitbank.html, accessed April 2007

Castlerock Estates, Inc. vs Estate of Markham (1994) 871 F. Supp. 360, US District Court for the Northern District of California

Connecticut General Statutes (2007) 47-42c. Acquisitions of restrictions

Draper, A. E. (2004) 'Conservation easements: Now more than ever – overcoming obstacles to protect private lands', *Environmental Law*, vol 34, pp247–282

Environmental Law Institute (2002) *Banks and Fees: The Status of Off-Site Wetland Mitigation in the United States*, Environmental Law Institute, Washington DC

Florida Department of Protection (2005) *State Secures Land for Preservation of Wekiva River Basin*, www.dep.state.fl.us/secretary/news/2005/07/0726_01.htm, accessed April 2007

Fox, J. et al (2006) 'Conservation banking', in Scott, M., Goble, D. D. and Davis, F. W. (eds) *The Endangered Species Act at Thirty: Conserving Biodiversity in Human-Dominated Landscapes*, Island Press, Washington DC

Gardner, R. C. (1991) 'Public participation and wetland regulation', *UCLA Journal of Environmental Law and Policy*, vol 10, pp1–39

Gardner, R. C. (1996) 'Banking on entrepreneurs: Wetlands, mitigation banking, and takings', *Iowa Law Review*, vol 81, pp527–587

Gardner, R. C. and Radwan, T. J. P. (2005) 'What happens when a wetland mitigation bank goes bankrupt?' *Environmental Law Reporter*, vol 35, pp10590–10604

Gardner, R. C. and Radwan, T. J. P. (2006) 'What happens when a wetland mitigation bank goes bankrupt?' *National Wetlands Newsletter*, vol 28, issue 4, p1

Hawaii Revised Statutes (2006) 195D-2, Conservation of Aquatic Life, Wildlife, and Land Plants

IMDb (2003) 'Memorable Quotes for Pirates of the Caribbean: The Curse of the Black Pearl', http://www.imdb.com/title/tt0325980/quotes, accessed April 2007

In re The Environmental Trust, Inc. (2005) Combined Disclosure Statement and Liquidating Plan of Reorganization, dated as of 20 December 2005, No. 05-02321-LA11, US Bankruptcy Court for the Southern District of California

Kelo vs City of New London (2005) 545 US 469, US Supreme Court

Klass, A. B. (2006) 'Adverse possession and conservation: Expanding traditional notions of use and possession', *University of Colorado Law Review*, vol 77, pp283–333

Loretto vs Teleprompter Manhattan CATV Corp. (1982) 458 US 419, US Supreme Court

Lucas vs South Carolina Coastal Council (1992) 505 US 1003, US Supreme Court

Mahoney, J. D. (2002) 'Perpetual restrictions on land and the problem of the future', *Virginia Law Review*, vol 88, pp739–787

Mann, C. C. and Plummer, M. L. (1995) *Noah's Choice: The Future of Endangered Species*, Alfred A. Knopf, New York

National Mitigation Banking Association vs US Army Corps of Engineers (2007) No. 06-cv-2820, 2007 WL 495245, US District Court for the Northern District of Illinois.

O'Neil vs Picillo (1989) 883 F.2d 176, US Court of Appeals for the First Circuit

Palazzolo vs Rhode Island (2001) 533 US 606, US Supreme Court

Penn Central Transp. Co. vs New York (1978) 438 US 104, US Supreme Court

Pennsylvania Coal Co. vs Mahon (1922) 260 US 393, US Supreme Court

Reiland, R. (2004) *Shoot, Shovel & Shut Up*, www.lewrockwell.com/orig4/reiland3.html, accessed in April 2007

Robertson, M. M. (2006) 'Emerging ecosystem service markets: Trends in a decade of entrepreneurial wetland banking', *Frontiers in Ecology*, vol 4, pp297–302

Ruhl, J. B. (2004) 'Endangered Species Act innovations in the post-Babbittonian era: Are there any?', *Duke Environmental Law and Policy Forum*, vol 14, pp419–439

Tennessee Valley Authority vs Hill (1978) 437 US 153, US Supreme Court

United States Code of Federal Regulation (2007) 26:1.170A-14. Qualified conservation contributions

United States Code (2007) 5:553. Administrative Procedure Act

United States Code (2007) 42:9607. Comprehensive Environmental Response, Compensation, and Liability Act

United States Code of Federal Regulations (2007) 50:17.3. Endangered and Threatened Wildlife and Plants

United States Code (2007) 16:1531-1544. Endangered Species Act

United States Fish and Wildlife Service (2003) *Guidance for the Establishment, Use, and Operation of Conservation Banks*, http://www.fws.gov/endangered/policies/conservation-banking.pdf, accessed April 2007

United States Fish and Wildlife Service and California Department of Fish and Game (1996) *Supplemental Policy Regarding Conservation Banks Within the NCCP Area of Southern California*, www.dfg.ca.gov/hcpb/conplan/mitbank/mitbank_policies/fullsuplmnt.pdf, accessed April 2007

United States vs Hawkins (2005) Amended Complaint, Civil No. 3:05CV-12-H, US District Court for the Western District of Kentucky

United States vs Hawkins (2006) Consent Judgment, Civil No. 3:05CV-12-H, US District Court for the Western District of Kentucky

United States Public Law (1998) No. 105-178, Transportation Equity Act for the 21st Century

Washington Revised Code Annotated (2007) 47:12.330, Advanced environmental mitigation

Western Prop. Serv. Corp. vs Shell Oil Co. (2004) 358 F.3d 678, US Court of Appeals for the Ninth Circuit

Wilkinson, J. and Thompson, J. (2006) *2005 Status Report on Compensatory Mitigation in the United States*, Environmental Law Institute, Washington DC

7

Regulatory Considerations

Susan Hill

Introduction

Conservation banks are established under a cooperative strategy that assigns a monetary value to conserved habitat, harnesses market forces, and simplifies and streamlines the regulatory process. Through their size, habitat quality and level of protection, banks improve significantly upon traditional ways of conserving habitat necessary for species recovery and survival. Through partnerships with private landowners, conservation banks are one tool the US Fish and Wildlife Service (USFWS) can use to increase available habitat for protected species, in order to help fulfil the USFWS mission and comply with the mandates of the Endangered Species Act (ESA), while benefiting the private landowner. Although USFWS focuses on the ecology and protection of species in conservation banks, which is the focus of this chapter, and it does not directly regulate the economic aspects, the positive economic benefits to landowners are clear. Conservation banking is a free market enterprise that allows individuals and entities to profit from the protection of species through credit sales at values determined by the marketplace. At the same time, it provides a simple, direct method for applicants to compensate for impacts to the same species. Therefore, it is not only the protected species that benefit from conservation banks, but also the bank owners and the project applicants, resulting in a win–win–win situation.

Government agencies, such as the California Department of Transportation, and commercial industries, such as Pacific Gas and Electric Company, have also recognized the value of establishing their own banks, which they can use for themselves, and to sell excess credits for a profit. Not only do the banks provide a streamlined form of compensation for projects, as well as a source of supplemental income, but the public relations component adds another valuable dimension to the effort. The public generally likes to know that our agencies and industries are 'doing the right thing' for the environment, and banks provide an opportunity to show the public many acres of healthy habitat that companies have protected in perpetuity.

In order to explain the context within which conservation banks have developed, this discussion begins with a very cursory overview of the regulatory umbrella of the ESA of 1973, as amended, that USFWS helps to administer. Subsequent topics herein describe decisional factors to consider when evaluating the potential of a certain property or location for a conservation bank, the primary regulatory elements of a conservation bank, and a brief peek at current trends in conservation banking.

This information is written based on the practices of the Sacramento Fish and Wildlife Office (SFWO) of USFWS. The majority of conservation banks are located in California and many of these have been processed through the Sacramento office. Although other field offices of USFWS support conservation banking, and the underlying principles are the same for all, individual offices may have some variations as to certain practices and policies. A potential bank operator should always check with their local field office about any policies described in this article prior to beginning a bank project. The bank operator should also keep in mind that other agencies are frequently involved, especially in California, where the US Army Corps of Engineers (USACE), a primary agency in mitigation (wetland) banks and the California Department of Fish and Game (CDFG) also play active roles in both mitigation and conservation banking. The agencies work together as a team to establish the banks; however, each agency has its own contributions, issues and concerns to address. For instance, one example of different processing due to agencies collaborating is that when USACE is not involved in a bank (any bank without jurisdictional wetlands), USFWS and the State of California use a conservation bank agreement as the main contract to effectuate a bank for species only, whereas when USACE is involved with the other agencies, a bank enabling instrument is the primary contract used to incorporate the wetlands. This concept is further explained in the section herein on Processing the Essential Documents.

Conservation banks are designed to continue 'in perpetuity' – or forever. When developing banks, we all must function under the constraints of our abilities to predict with any degree of certainty what will happen 10, 20, 50 or 100+ years from now. Recognizing the uncertainty of these predictions, especially in the face of climate change, the goal is always to keep the concept of perpetuity in the forefront and do the best we can with the knowledge we have now – incorporating the best available science with reasonable allowances for adaptation to changing circumstances.

Overview of USFWS regulatory framework as applied to conservation banks

There is no specific USFWS regulatory authority for conservation banking (see Chapter 6, Legal Considerations). Rather, it has developed as a tool to carry out the mandate and the mission of USFWS to conserve species. This basic overview of how conservation banking fits into the regulatory scheme forms a backdrop for understanding USFWS purpose and function in conservation banking.

The mission statement for USFWS is: 'Working with others to conserve, protect and enhance fish, wildlife, and plants and their habitats for the continuing benefit of the American people. Under the ESA, '[t]he terms "conserve", "conserving", and "conservation" mean to use and the use of all methods and procedures which are necessary to bring any endangered species or threatened species to the point at which the measures provided pursuant to this chapter are no longer necessary.' Conservation banking is a collaborative, incentive-based approach to land conservation that can assist substantially in fulfilling the recovery of imperilled species while enriching the lives of private landowners. This concept has proven to be a successful mechanism for upholding the USFWS mission.

The ESA applies to all federal departments and agencies, and is administered in part by USFWS. The policy of the ESA, under section 2(c)(1), states that '[i]t is further declared to be the policy of Congress that all Federal departments and agencies shall seek to conserve endangered species and threatened species and shall utilize their authorities in furtherance of the purposes of this chapter.' The purposes of the ESA, as stated in section 2(b) therein, include 'to provide a means whereby the ecosystems upon which endangered species and threatened species depend may be conserved, [and] to provide a program for the conservation of such endangered species and threatened species…' Conservation banking supports this ESA policy and its purposes beautifully, as banks are based upon the preservation of self-sustaining ecosystems for the benefit of species protected under the ESA. In fact, one of the evaluating criteria of any prospective bank is the land's ability to maintain its habitat qualities in perpetuity, with little to no human intervention.

USFWS evaluates various factors, as described in the ESA, to make determinations whether a species is an endangered or a threatened species based on the best available science. Species identified as being in danger of extinction or likely to become so within the foreseeable future will be 'listed' as endangered or threatened, and species that have recovered sufficiently may be 'de-listed', as described in the ESA, section 4. As a part of attaining de-listing of species, the ultimate goal of the ESA, USFWS is charged with developing and implementing 'recovery plans', when it is found that such a plan promotes the conservation and survival of listed species.

Each recovery plan should incorporate site-specific management actions as may be necessary to achieve the plan's goal for conservation and survival of the species, as well as objective, measurable criteria that, when met, will allow the species to be de-listed. These criteria generally include the management and protection in perpetuity of suitable habitats within core areas for the species. This is to assure that the ecological integrity of the protected core areas are not threatened by adverse anthropogenic habitat modification. Conservation banks provide the management and the financial means to conserve species on the bank property, and protect the land in perpetuity, thereby playing an important role in the recovery of species.

Finally, sections 7 and 10 of the ESA require USFWS to consult with anyone who may impact federally protected species inhabiting a project area or action area. When USFWS finds through the consultation process that impacts to protected species will occur, USFWS may require measures to minimize impacts on the species. These minimization measures are within the terms and conditions of biological opinions provided by USFWS, and often include compensation. Compensation requirements are based on a number of factors, including the size and quality of the habitat in the action area and the listed species impacted. These requirements typically require conservation of a specific amount of land suitable as habitat for the species affected by the project. Often, the acreage required under a specific project is so small that, standing alone, it cannot provide the requisite perpetual protection. Conservation banks were developed in part to address this problem. In order for a piece of land to qualify as adequate to establish a bank, the acreage and qualities of the property must be such that the bank can be maintained as a self-sustaining ecosystem for the benefit of the species. Smaller pieces may then be sold as credits within this landscape, thereby avoiding the sustainability problems of too-small, scattered compensation sites.

Other regulatory benefits to both USFWS and the project applicant include the streamlining effect to the consultation process, and the cost-effectiveness, economy of scale, and time savings of implementing many compensation sites under one umbrella. Having compensation sites completely protected and ready to go at the moment that a project applicant is ready to break ground is a great benefit to both the project applicant and USFWS. Not only does it save the time and financial investment involved in putting an individual compensation package together, but it also prevents unwanted and expensive delays in trying to fulfil the required terms and conditions of the biological opinion prior to ground breaking. Further, it minimizes required inter-agency coordination to approve mitigation, since all of the agencies involved will typically be signatories to the bank agreement.

The decision process – evaluating a potential bank site

When a determination is made under the ESA section 7 or 10 that a project impacts a federally listed species, USFWS preferred alternative is commonly avoidance. The project should be designed to avoid impacts to listed species as much as possible by finding ways to build in areas that have no established habitat, or those with lesser value habitat, as much as possible. Preserving on site the best available established habitat should be the primary goal. The resulting coexistence of people and listed species is often a viable possibility, and may reduce the cost to the project applicant. At the same time, protecting pre-existing habitat avoids loss of the species through habitat destruction and direct take of the species; provides known functioning habitat because it is already occupied by the species; avoids the temporal loss that results when new habitat must be created and re-occupied to replace the lost habitat; and avoids the risks inherent in creating new habitat with the hope it will successfully support all of the affected species. However, it is not always feasible or practical to protect habitat on site. Or there may not be enough habitat on site available to compensate for the losses resulting from the project. Therefore, it is often necessary to implement some kind of off-site compensation, which is where conservation banking on an appropriate piece of land plays a key role.

Two primary options exist for off-site compensation – a project applicant can either purchase and protect an individual piece of property that will fulfil the terms and conditions of the biological opinion, often termed a 'turn-key' site, or purchase credits from an approved conservation bank that will fulfil the agency requirements. A turn-key compensation site is designed to fulfil the same terms and conditions of the biological opinion as purchasing credits at a bank, only turn-keys protect one project at a time, as opposed to a collection of pre-approved sites, as at a conservation bank. To establish the turn-key site, a piece of property of adequate size, quality and biological composition must be located, purchased and protected by the applicant in a manner very similar to the requirements for establishing a bank. The following information will be relevant to any turn-key sites as well to banks.

The decision to establish a conservation bank in any particular location, and how best to use the property, requires an evaluation of a number of factors by the agencies and the prospective bank owner.

The basic considerations

Location, location, location!

Banks should be strategically located in order to best serve the species, and to integrate with the needs of urban development. USFWS looks to its recovery plans and to critical habitat areas, whenever they are available, for prime areas in need of protection that fit into the overall strategy to recover species. Recovery plans are obtainable from USFWS by the general public. However, many properties outside of these designated areas, or properties benefiting listed species that do not yet have a recovery plan, may still qualify as beneficial and worthy of protection, and should always be fully evaluated for their merits.

Are there any federally listed species on site? If not, is there appropriate species habitat on site, and some documented sightings of species in the nearby vicinity that could inhabit and benefit from conserving or restoring the proposed property? Occasionally a proposed bank site may not currently support species, but fits well within a regional recovery plan for one or more species, and can reasonably be anticipated to support the species in the future, either by preservation or restoration. The site must be large enough, either on its own, or as a property adjacent to some other fully protected property, to be capable of sustaining habitat with minimal human intervention. This would include its own water supply and hydrological functioning, food sources for the species to be protected, shelter and, often, breeding areas for the species for their ultimate survival.

Which properties are the most desirable?

When evaluating which properties will be suitable for conservation banking, regardless of a location within or outside of a predetermined recovery area, there are a number of factors to consider as to its ecological fitness and sustainability. None of the factors alone is necessarily determinative; they are to be considered in the totality of the circumstances.

Conservation biology factors

As previously mentioned, parcel size is considered primarily as it relates to the self-sustainability of the habitat. In evaluating the self-sustainability, some of the other contributing factors to look at include the shape of the parcel, the present and future edge effects, and watershed integrity and defensibility.

Consider edge effects that occur as a result of the surrounding land uses. Agricultural land uses may be friendlier in the way of open space, with less human impact, but may also result in contaminated run-off from irrigation waters, and overspray of pesticides or other agricultural chemicals. The land must be adequately defensible from external adverse impacts to provide protection for the credited habitat and species. Different species will have varying sensitivities to these chemical impacts that must be taken into consideration. And though a property may presently be farmed, how likely is it that it will remain that way? If adjacent lands are already developed into housing or commercial property, or consist of open space perhaps waiting to be developed, how much will future anthropogenic impact affect the bank property? People will use beautiful open spaces for a variety of incompatible activities such as walking dogs, letting cats wander about, all-terrain vehicle riding, hunting, shooting ranges, camping, trapping, etc. How well can the habitat and the species be protected from these potential impacts?

Impacts due to edge effects will be minimized where there is less edge to impact. This occurs when a parcel has a round, smooth edge, or square, straight edges. Bank properties with areas that may be isolated or protruding from the main bank body, such as parcels with multiple finger-like peninsulas and zig-zagging boundaries, may be impacted more heavily by activities outside the bank property because of the larger surface area potentially affected.

Finally, when evaluating watershed integrity and defensibility, depending on the species and habitat, the relative importance can vary considerably. For a vernal pool bank, protecting plants and crustaceans endemic to the pools, the microtopographic watershed is a much more developed and intricate system to consider than for a San Joaquin kit fox, which will simply need water for survival. In a vernal pool or similar type bank, the frequency and duration of ponding and soil saturation must be maintained.

Also important hydrologically and ecologically is the ratio of uplands to wetlands. The density of vernal pools or other wetlands should not exceed that of local reference areas, or the historical density for that particular location. The USFWS biologist looks to see that there is sufficient non-wetted area available for the upland species, such as salamanders during their aestivation, or the solitary bees that pollinate many vernal pool plants.

Land use and the current condition of the property

If the land is not in its original, pristine condition, what were the prior land uses? What is the current use? What degree of physical disturbance has taken place? To what extent have exotic species moved in and pushed out the native species? Have the prior and current land uses impacted the land to the point that it needs restoration to function as habitat? Does the cost–benefit ratio of restoration to bank profits favour the bank owner, and thus the bank's viability? On the other side of the coin, native species diversity and richness, as well as habitat diversity, are to be considered for both habitat quality and the marketability of the bank. Having a larger pool of species will make it faster and easier for the banker to sell credits. Marketability is important to USFWS as well as to the bank owner, as USFWS wants the banker to be successful in providing the required compensation, and to benefit from doing so.

Regional land use

Another important factor to consider is the proposed bank property's contribution to known species corridors, and its adjacency to other protected lands. Species migration through interconnectivity of habitat areas is important for genetic dispersal and diversity, and reduction of the impacts of localized habitat losses. Proposed bank properties that lie within important habitat corridors are greatly valued. Lands that are directly adjacent to already-protected lands are also highly prized for their overall contribution to the bigger picture. A proposed bank property that is adjacent to a state park, or other protected habitat area, may also be acceptable as a much smaller parcel of land than that required for a stand-alone bank because of its ability to piggyback onto the existing conservation values of the other property.

Service area

The service area is a geographic area assigned to the conservation bank within which credits may be sold to compensate for project impacts. Service area is based primarily upon ecological factors, such as the genetic distribution of a species, watershed configuration, soil attributes, etc. Adverse effects on a specific listed species may only be compensated for in a bank that supports the species being affected. A bank may have more than one service area because different species may have different service areas based upon their specific ecological attributes. Map(s) and a physical description of the bank's service area(s) are included in the banking documents, and also are provided by USFWS on our conservation bank website, along with contact information, types of credits approved for sale, and the size and location of the banks.

The bank's service area will provide the primary market base for the bank's credit sales. Bankers need to evaluate the cost-effectiveness and the marketability of their potential bank based on the market base available to them within the particular service area. Because conservation banking is a free market enterprise, the banker may charge what the market will bear; however, the banker is responsible for making sure the market base is there to support the bank. A few good resources to consult for proposed future impacts and development within an area could include the local branches of the various resource agencies, local planning commissions, chambers of commerce and general plans for the area.

Occasionally, a project outside of a bank's service area will be allowed to purchase at that bank, but this is approved on a case-by-case basis. Where service areas of different banks overlap, a project applicant will have the option of choosing which bank to use. Because it limits bank monopolies and extreme credit prices, USFWS encourages this type of competition between banks.

What is the best use of the land as a conservation bank?

The decision to preserve, restore or create habitat on a property will primarily depend on the condition of the land, the supported habitats and species, and what types of credits may be sold.

Types of credits

There are essentially two basic types of credits available at conservation banks – 'preservation credits' and 'creation credits'. Preservation credits are the simplest form of credit to establish. They comprise habitat that is already existing and functioning on the ground. Species surveys are conducted, and a baseline report is compiled by a qualified biologist describing all of the relevant ecological conditions on the proposed bank property. From this information and a site visit by a USFWS biologist, as well as a legal map and legal description of the proposed bank site, a determination can be made as to the number of preservation credits to award for each species. If there are some degraded areas on the property, while others areas are in relatively good condition, preservation credits may be awarded for the existing habitat that currently supports the species, with a tentative agreement as to an increase in credits when remediation actions have been implemented to the degraded areas, restoring their functionality. These areas may also qualify as creation credits, depending on their overall attributes – such as, what species they are for, how much work is required to restore them to functioning status, etc. Preservation credits are typically acknowledged for sale upon bank establishment or opening; creation credits will be awarded in increments as performance criteria are met and the requisite habitat is successfully established.

Creation credits are a bit more variable as to definition, and require varying amounts of greater effort and upfront investment to implement; however, they are much needed for certain species, and in high demand, due to their scarcity. In particular, compensation for vernal pool impacts has a fairly consistent creation component as well as a preservation component. 'Creation' of habitat can vary along a sliding scale, under USFWS interpretation, to include enhancement, restoration or full creation of the requisite habitat. Proposals for creation credits will be reviewed closely to ensure that the plan appears to be practicable and has a very good chance for success.

Enhancement entails the least degree of change from the current condition of the habitat in order to renew the habitat so that it better supports the listed species. This may require, for example, no more than planting some native plants that will provide increased breeding, feeding or sheltering habitat for the targeted listed species in an area that has suffered from some damage such as fire or sustained flooding.

Restoration entails a higher level of effort to reinstate the habitat. For example, in an area where the natural land topography was once very hilly and included vernal pools and swales, but has since been levelled, disked and used for agricultural purposes, restoration is much more extensive. Here, the banker would need to look to historical and current aerial photos to determine the previous footprints of the pools, and attempt to restore the land to its prior condition to the greatest extent possible. The process is much more complicated and involved than it appears from this simple description; however, the science is developing in this field to better accomplish satisfactory results.

Full creation of habitat occurs when habitat is created where it has not previously occurred. Generally speaking, this type of 'creation' is the least desirable, and the most difficult to successfully implement. Habitats develop naturally in certain areas because the environmental factors are amenable to those habitats. Attempting to insert a particular type of flora and fauna where it has not previously existed may be like putting the proverbial round peg in the square hole. It just doesn't fit, and if you force it, it never quite looks right, or functions properly. Vernal pool habitat is very rarely, if ever, amenable to this form of creation due to the very specialized soil requirements, and the very specific, endemic species, and their habitat requirements. One area in which this form of creation of habitat has been successful, albeit in a sliding scale region between restoration and full creation, is in providing habitat for California's valley elderberry longhorn beetle (VELB).

VELB habitat consists of elderberry shrubs existing in a mixed canopy of taller and shorter associated native plants, often in riparian areas (along the edges of streams, creeks and rivers). If a property can be shown to be highly likely to support the ecological requirements of VELB habitat, and has functional VELB habitat along the same waterway, or otherwise nearby, then creation of VELB habitat in that area may be a viable option, even if there is no record of elderberry plants in that particular location in the past. Soils, hydrology and other important factors should be carefully evaluated, and approval by USFWS must be obtained, prior to beginning any creation project.

Performance objectives are required for any creation crediting, and must be successfully met in order for credits to be released for sale. Typically, credits are released for sale in an incremental release fashion, where, as each individual objective or set of objectives is met, a portion of available credits is released. This occurs successively over a period of time until all of the credits have been released, and the habitat has proven to be fully functional. If problems do arise along the way, remediation and/or alternative plans may need to be implemented. USFWS will work closely with the banker to correct the problems, or to find an alternative option.

Crediting methods

In the past, SFWO has used special credit determination methods for vernal pool and California red-legged frog credits, but these special valuations are no longer relevant, partly because of growth and changes in the banking industry. The primary purpose served by the credit determination methods and the resulting 'bank value' multiplier (greater than 1) was to increase the number of credits available per acre at banks due to the fact that the majority of impact sites that were being compensated for at the banks were of very low value habitat and of very small size, less than one acre each. Conservation banking has grown substantially since these credit determination methods were developed, and many more projects of significant size, well beyond one acre, impacting high habitat value areas, are turning to banks for their compensation. In order to avoid confusion and the messy calculations required to convert bank 'credits' to 'acres' for these large-scale projects, SFWO has decided to issue vernal pool and red-legged frog credits on a consistent basis for all banks and for all projects. The issue of low value, small impacts versus high value, large impacts, and the requisite compensation required for each, will now be addressed strictly through the conservation ratios applied during consultation with USFWS. This means that when the biological opinion or habitat conservation plan is written, the size and quality of the impacted habitats will be factored into the decision as to how much habitat should be provided as compensation.

In its simplest form, one credit equals one acre. In many of our upland species, such as the San Joaquin kit fox, this is the case; and for wetlands, the only variation is that one 'wetted' (or wetland) acre equals one credit. If only it remained that simple, crediting would be so easy. However, it is imperative that the method used to determine the impacts to a species for a biological opinion is the same method used when evaluating the number of credits to be issued for a particular species at a bank, and different species have varying requirements that call for alternative evaluations. For instance, in the case of VELB, project impacts are evaluated under a programmatic agreement that describes impacts and their respective compensation based on an 1800 square-foot unit. In keeping with the primary consideration of crediting a bank commensurately with the assessment of the impacts to a project site, for VELB described above, one credit currently equals 1800 square feet of VELB habitat. In the case of a delta smelt (fish) bank, its habitat is along the shore of the waterway, in areas shaded by aquatic plants. In this case, the crediting scheme is based on linear feet of shaded riverine aquatic habitat. A bank operator should be sure to discuss the method used to determine credits for the particular species on the proposed property with the local field office.

Dual and multi-species credit issues arise when more than one species share the same habitat on a bank property. When this occurs, the overlapping areas, shared by all species, may be sold to compensate for any one, or any combination of the species sharing the habitat. However, once the credit is sold for any species or combination, it is removed from the total bank credits, and may not be sold for other species, not included in the first sale. For instance, if a vernal pool credit may be sold for three different types of crustaceans that live in the pool, and it is only sold for two of the species impacted at a project, it may not later be sold for the third species. The credit is gone from the bank's inventory, because the habitat acreage has already been sold. Selling the same approved area twice is considered 'double-dipping' (see Chapter 11: Credit Stacking). Multi-species credits can lead to some tricky credit sale tracking that is best handled by someone good with mathematical skills and spreadsheets.

Processing the essential documents

A conservation bank has two or three main components – the basic two being the contractual element, which describes the responsibilities and obligations required of the banker in exchange for the ability to sell credits, and the management element for the perpetual conservation of the land and stewardship of the habitat and species being protected. Where there is a creation component, a third element, the development plan, is also required to describe and implement the restoration/creation of that portion of the bank habitat. The process of establishing a conservation bank includes gathering the relevant information pertaining to the species to be protected, and to the land to be conserved, and assembling this information into a useable format for use in establishing and managing the bank property in perpetuity. After a brief history of the overall, ongoing process in California, this section will introduce the reader to some of the fundamental and supporting documents required to execute these two elements.

At the end of this section is a list of information that USFWS requests in order to make a preliminary determination as to the potential of a particular property as a conservation bank.

California conservation banking – a short history lesson

The banking process is not always as clear-cut and smooth as we would all like for it to be, and has experienced its share of growing pains. From its simplistic beginnings, conservation banking grew incredibly fast in popularity, and quickly outgrew its humble roots. In California, as the first group of banks began maturing, problems were discovered in these 'prototype' banks that had to be worked out in order to avoid making the same mistakes again. At the same time that these bugs had to be corrected, applications for new banks were pouring in, agency staff and potential bankers each had their own conflicting ideas as to how to make the programme better, and the budget and the staff for the banking programme was shrinking. This unfortunate series of events resulted in a backlog that became quite frustrating for the potential bankers and the agencies. Recently, SFWO has increased the number of staff assigned to the conservation banking programme, and there has been significantly more coordination among the resource agencies to coordinate and resolve the early issues, with an emphasis on standardizing the requisite bank documents.

Many factors contribute to the overall time involved in establishing a conservation bank: variations in the timing of surveys for different species; a variety of documents often requiring several supporting documents and iterations; differing concepts of how contract terms should be written by different potential bankers; and the coordination of numerous government agencies. The resource agencies in California are currently working very hard to coordinate a number of templates and model documents in advance in order to streamline the processing of banks. Templates have worked well in the past to streamline processing, when they were used as written. However, when significant changes are made in templates for individual banks, as was occurring frequently, review time increases dramatically. By publishing the proposed templates in the Federal Register for public review and comment, the agencies are committed to making these templates as concrete as possible, so that modifications will be minimized, and processing speed improved. Previous templates were not always coordinated through all agencies, resulting in possible changes that would have to be re-negotiated later. These current efforts remedy that problem as well. Hopefully, the templates developed from these concerted efforts will be beneficial to all who are interested in establishing a conservation bank, whether or not it is in California.

Essential conservation bank documents

There are four primary components that form the heart of a conservation bank: the conservation bank agreement (CBA) or mitigation bank enabling instrument (BEI) (see Appendix II); conservation easement (see Appendix III); management plan/ development plan; and the financial assurances. Some of these elements require various supporting documents to make the picture complete, or to 'flesh out' the basic information. Following is an overview of the information, and some of the key elements, that USFWS looks for in order to approve a conservation bank.

Conservation bank agreement/mitigation bank enabling instrument

A CBA is the template used by USFWS when the bank is proposed for species protection only, and does not involve waters of the US that are under the jurisdictional authority of USACE. See Appendix II for a basic template developed for banks in California. When USACE is involved in the bank, and both species and wetlands are protected and included in the credits to be sold at the bank, then a BEI template is used. Both of these documents cover the same basic information and concepts, and are referred to here as the 'agreement' for simplicity. Both templates may be received by requesting them from the respective agencies.

The agreement is the overarching contract that fully describes the terms agreed upon by the agencies and those who operate the conservation bank. This contract includes the legal components of bank establishment, operation and management. It also describes general information such as jurisdictional authorities of the agencies; definitions of terms used in banking; the various informational exhibits required, which are incorporated by reference into the agreement; financial assurances; how credits will be released, sold and tracked; transfer or sale of the bank; and dispute resolution.

By limiting the agreement to general terms and conditions, it can be written as a general template document applicable to all banks, reducing review time substantially. All of the exhibits attached to the agreement are incorporated into the agreement by reference, and give the details that are specific to each particular bank.

The exhibits that are attached and incorporated into the agreement include all of the following, as well as other supporting documents. The three components described below contain the vital information to maintain banks in perpetuity. While the agreement remains in effect while credit sales are ongoing, eventually the agreement will no longer be required, and will terminate as a contract document. When the agreement terminates, the conservation easement, management plan and financial assurances are the three components that must be tied together and carry on the management of the bank in perpetuity in order to fulfil the conservation and recovery goals of USFWS.

Conservation easement

The conservation easement plays a most vital role in the protection of the land in perpetuity (see Appendix III for a template). The 'grantor' (owner of the property) grants an easement to a qualified 'grantee', transferring certain property rights to the grantee to hold for conservation purposes. The grantee must be a separate entity from the grantor, and is responsible for monitoring and enforcing the terms of the conservation easement to protect the conservation values for use as a species and habitat bank. The completed, signed easement is officially recorded at the County Recorder's office in the county where the property is located so that it will follow the title from one owner to the next, ensuring that each successive owner of the property understands that the property comprises a conservation bank and must be managed accordingly.

The conservation easement places certain restrictions on uses of the property in order to preserve the conservation values. The restrictions typically limit the property uses to personal uses by the property owner, such as grazing and various forms of private recreational activities. The conservation easement also extinguishes future development rights on the property, in order to protect it for the benefit of the species and habitat for which the credits are being sold. The property owner/ grantor may be eligible for significant tax benefits for recording a conservation easement, and should consult with their tax adviser about such a benefit.

In order to qualify to hold a conservation easement for a conservation bank, a grantee must be either a state government agency authorized to acquire and hold title to real property for the purpose of conservation of natural lands, or a qualifying non-profit agency. A non-profit agency must qualify by serving a conservation purpose under the Internal Revenue Code sections 170(h) and 501(c)(3), and the California Civil Code section 815.3. Compliance with these code sections is required to ensure that the grantee of the conservation easement is overseen by the Secretary of State, and has as its mission the conservation of natural resources, in order to best serve the purposes of the conservation easement and the conservation bank.

Certain preliminary information and documentation is required in order to ensure that a particular piece of land is appropriate to place under a conservation easement for acceptance as a bank by USFWS. First of all, the property owner must obtain a title report from a recognized title company, listing any and all liens and encumbrances to title. Each of the referenced lien and encumbrance documents must be obtained, and individually reviewed and evaluated by the property owner or her agent to determine whether they will affect the conservation values of the property. This written evaluation of each lien and encumbrance forms the 'property assessment', which becomes an exhibit to the agreement.

All monetary liens must be subordinated to the conservation easement by a subordination agreement between the property owner and the lender so that any default will not affect the conservation easement. The subordination agreement is recorded with the easement. Any easements on title which may subsequently affect conservation values of the bank property must be excluded from the total bank property acreage. The property owner then warrants by signature the total acreage that is unaffected by any liens or encumbrances, and available for protection consistent with the conservation purposes of the conservation easement, and that the property assessment is complete and correct to the best of their knowledge.

One issue that may arise in the transfer of bank property is a conflict due to the legal doctrine of merger. Merger occurs when the grantee of a conservation easement is the same party as the fee title owner of the property. If both roles are filled by the same party, a 'merging' of the fee title and easement occurs, and the conservation easement terminates. Such a situation must be carefully considered by all agencies prior to approval of the transfer, to ensure that the conservation of the bank property will not be compromised. Care must be taken that language is inserted into the conservation easement such that if a merger does occur, and the conservation easement ceases to exist due to a fee title transfer, the property will continue to be protected. The conservation easement should state that the terms of the conservation easement shall survive any merger of the fee title and the easement, perhaps through recordation of a covenant stating the terms, and, if and when the title is transferred back out of the merger situation, a conservation easement shall be reinstated simultaneously with the transfer to secure protection of the property in perpetuity.

Management plans

There are three main types of management plan that may be incorporated into any given conservation bank – development plans, interim management plans and long-term management plans. The development plans and interim management plans are designed to accomplish short-term results. The long-term management plan continues in perpetuity; is more susceptible to modification in order to accommodate changes in the environment or adjust activities that were unsuccessful for any reason; contains long-term provisions for monitoring, reporting and funding; and is the primary focus here.

Development plans describe the short-term, start-up plan for creating or restoring habitat on site, and include management tasks required to meet performance criteria in order to receive full credit release. Performance criteria are particularly important components in the start-up periods of banks where habitat is being created or restored. When success of the created habitat does not meet the standards set by the performance criteria, remediation may be necessary to bring the bank into compliance and release all potential credits.

Interim management plans are written to address the initial short-term management period while the endowment account is being fully funded, and the credits are being sold. Adaptive management could play a role if certain anticipated results do not occur, or if unexpected results do. However, the short time period consisting of only the first few years after bank establishment are typically quite successful, especially for preservation-only banks, and operations tend to be fairly smooth and predictable.

The long-term management plan, referred to here as simply the 'management plan', describes baseline information for the property, including topography, hydrology, soils, vegetation and wildlife on the property; management activities, objectives and goals to maintain or improve the habitat over time; and administrative information regarding funding, monitoring and reporting. It is the second key component to the success of the bank in perpetuity, and is written to address all management responsibilities as fully as possible, but with the understanding that adaptive management is key to ensuring long-term success. Adaptive management allows for changes in management practices as necessary according to evaluation of the information provided in monitoring reports, and observations from site visits.

Management activities are based on continuing to meet the performance criteria and objectives of managing the land for the benefit of the species being protected. Tasks will include physical maintenance of the property such as fencing and signage to discourage trespassers and dumping, and to define the bank boundaries. Biological management includes such activities as weed abatement through manual removal, spraying or fire; thatch reduction and weed abatement through grazing management; providing strategically located water troughs for livestock to reduce impacts to wetlands; and invasive species control methods, such as draining stock ponds annually to reduce the impact of exotic bullfrogs. This section should be written by an experienced biologist who understands the ecology of the local area, and the needs of the species being provided for by the management.

Administrative tasks described in the management plan include financial assurances, monitoring requirements and annual reporting obligations. Financial assurances may include both initial funding securities, and long-term endowment funds for perpetual management. All of the initial expenses are the responsibility of the bank owner, and are insured through securities. The long-term management obligations, including the monitoring and reporting duties, are funded through the long-term endowment so that there is no personal responsibility to provide funds, only to ensure that the tasks are completed. These assurances warrant specific attention, and are discussed more fully in the following section.

Monitoring responsibilities are determined on a case-by-case basis according to the type of monitoring, and the habitats and species being protected. Species monitoring at USFWS protocol survey level, or a reduced survey level, is usually only required for baseline reports, and then the period between such surveys will lengthen as long as no problems arise. The first few monitoring years will typically establish a baseline for future performance levels. A common schedule for vernal pool protocol level monitoring is for years 1, 2, 3, 5, 7, 10, 15, 20 and every 10 years thereafter, as long as no problems arise. Should any adaptive management or remediation be required to adjust habitat performance or management strategies, monitoring may increase in frequency for a few years until the area is stabilized once again.

Reporting is required annually, but the contents of the annual report will vary from year to year. Specific reporting requirements are negotiated according to site-specific demands, and so the information presented here is necessarily general; however, it provides a good guide as to the purposes being served by the monitoring and reporting.

One element that is required annually is a report on the status of the conservation easement. The grantee of the easement must inspect the bank property and file a report one or more times per year, depending on the location and the likelihood of trespassing or other adverse impacts. The easement report should contain information about the condition of the property generally, such as the condition of fences and signs; any impacts from trash, dumping, off-road vehicle riding or other trespassing; or buildings, roads or other prohibited structures built on site.

Biological reporting includes either full surveys, in the required years, or a spot-check type of inspection. As described above, the full, protocol level surveys are used to establish a baseline, demonstrate initial success for a few years after bank opening, and give an updated status report on a periodic basis after a few successful years have passed. The spot check surveys are done in the remaining years. A qualified biologist should visit the bank property at the appropriate time of year, and check the known locations of the species protected on site to see how they are faring for the year. If there is grazing on the bank property, annual surveys of the impacts of grazing are also required, in order to prevent over- or under-grazing, and any resulting adverse impacts.

Financial reporting is also required annually. Information regarding the annual endowment fund principal, interest and calculations for annual adjustments for inflation should be included in the report. Also, any income from credit sales, grazing or other source that is included in the endowment fund should be shown, as well as any and all expenses incurred in the bank's operations and management.

Financial assurances and endowment funds

While a complete discussion of the requirements of financial securities and appropriate endowment funding is the subject of an entire chapter (see Chapter 9: Financial Considerations), a basic overview is discussed here. There are two primary types of financial assurances used in conservation banking – securities and endowments.

The banker is responsible for all initial funding requirements of the bank out of pocket, until the long-term endowment is fully funded and available for use. Securities are required in the initial stages of conservation bank development and establishment to ensure successful implementation of creation and restoration plans, and to insure that all management needs will be fulfilled prior to full funding and use of the long-term endowment funds. The securities are used by the agencies for payment only as a last resort. If the securities are used for any reason, the terms of the agreement should require that the securities must be fully restored by the banker within a certain time period, usually a calendar year. If the securities are not used, they are refunded to the banker when they are no longer required as insurance.

The three possible securities are construction, performance and interim management securities. The form these securities take varies by agency, but may include letters of credit, bonds or cash. Their uses are fairly self-evident by their titles. The construction security is funded to assure that all promised construction will be completed; the performance security is held until performance criteria are met, to be used in the event that emergency funding for remediation becomes necessary; and the interim management security is an emergency reserve for bank management.

The primary funding instrument of a conservation bank is the long-term management endowment fund. The endowment is typically funded by depositing a percentage of each of the first credit sales until the entire amount is available. The amount to be set aside to endow a conservation bank depends on a number of factors, including both management costs and economic costs. Management costs are determined by a close analysis of the management plan. The total required to manage the bank is found by analysing the time and labour costs, the supply costs, and the periodicity of repetition of tasks required for each and every management task described in the management plan. Annual tasks include easement monitoring, biological surveying, reporting, etc. The annual management costs will be less than the total endowment revenue generated each year due to the tasks that are required periodically under a time period greater than one year, such as protocol level surveys, or fence replacement. The excess revenues should remain in the endowment fund to grow and generate interest until it is needed, and to generate the additional funds required to increase the principal amount to keep up with inflation. If any excess revenues are generated for any reason, that money may be used for unexpected contingencies, remediation, catastrophic events or other emergencies. However, the bank owner is not required to fund 'force majeure' events (catastrophic events) beyond any excesses available from the endowment account.

The basic concept of the endowment is to establish a fund and assign it to a qualified, non-profit trustee who will invest the endowment money under a balanced, non-wasting, reasonably secure economic portfolio. The endowment principal should not be allowed to dip below the original total deposit amount, as adjusted annually for inflation. A reasonably obtainable rate of return must be used to calculate the annual revenues that will be generated by the principal amount over the long term, and used for management activities. With such an account in place, the revenues may be used to manage, monitor and report on the conservation bank in perpetuity, for the benefit of the species for which it was established.

Trends in banking

Templates and models of several bank documents are currently being developed for California that should be useful to anyone considering or preparing a conservation bank. These guides will eliminate the need to 'reinvent the wheel' for each bank. The templates are being negotiated jointly with all involved agencies in California (National Oceanic and Atmospheric Administration, USFWS, CDFG, USACE) in order to streamline the entire review process, and both the draft BEI template and the draft conservation easement template were published in the Federal Register for public comment in 2006, and should be completed for public use in 2007. Any changes made to the final templates in individual bank applications may slow the process considerably, but when used in the manner developed, processing time should be reduced considerably from previous timetables.

The first joint templates to come out are a BEI template, a conservation easement template and a management plan outline. Status updates on these documents are posted on the SFWO website as they become available.

A list of requested documents to include in a bank proposal has been developed to aid in determining suitability of a proposed bank property prior to investing too much time or money. These documents will also help the agencies and the potential banker in the overall planning process. This information should be compiled into a bank proposal package for discussion and evaluation purposes at a meeting with the appropriate agencies.

Preliminary bank proposal

Recommended documents to include in a preliminary bank proposal to the agencies include the 14 items listed below. This information will allow the agency staff to further discuss the potential of the site as a bank, and to give an initial determination as to whether they believe it to be feasible or not, and what further information may be necessary prior to filing a draft of a complete bank document. A site visit for agency personnel to see the resources first hand is also recommended at the earliest opportunity.

1 ***Proposed bank name** – a short name based on a geographic feature if possible.
2 ***Bank contacts** – include the name, address, phone, fax, email and role in project.
3 ***General location map and address** of the proposed bank property.
4 ***Proposed bank boundaries** – accurate current map of the proposed bank property, showing boundaries of the bank.
5 ***Aerial photo** of the bank and surrounding properties (both current and historical).
6 ***Description of the site conditions** – include at minimum: current site conditions and habitats; photos of the site; description of wetlands and waters present on site; hydrology description; number of acres of existing wetlands and waters and what is proposed for creation, enhancement, etc.; site history including past land uses; and surrounding land uses and zoning along with the anticipated future development in the area.
7 **Biological resource survey** report.
8 **USACE-verified map of on-site jurisdictional wetlands and waters**, if required.
9 **Proposed number and type of credits** on the property.
10 **Proposed credit release schedule.**
11 ***Map of the proposed bank service area** for each type of credit proposed.
12 **Preliminary title report** indicating any easements or other encumbrances – provide a written assessment of all liens and encumbrances describing how each may affect bank operation or habitat values.
13 **Property assessment** – an affirmative statement that a conservation easement covering the property or fee title transfer of the property will occur as part of bank establishment. Include number of acres of 'functional' bank acreage based on exclusion of any easement areas that allow uses incompatible with conservation.
14 **Other** – describe any other known restrictions on the property.

Items marked with an asterisk may be submitted as a 'conceptual proposal', for those who would like a more preliminary screening and determination of agency approval to proceed prior to further investment of their resources.

Bibliography

California Civil Code, section 815 et seq, www.leginfo.ca.gov/calaw.html, accessed 15 May 2007

Department of Defense, Environmental Protection Agency, Department of Agriculture, Department of the Interior, Department of Commerce (1995) 'Federal guidance for the establishment, use and operation of mitigation banks', *Federal Register*, vol 60, pp58605–58614, 28 November

Internal Revenue Code, sections 170(h), www.fourmilab.ch/ustax/www/t26-A-1-B-VI-170.html, accessed 22 October 2007 and 501(c)(3), www.fourmilab.ch/ustax/www/t26-A-1-F-I-501.html#_6_3_, accessed 22 October 2007

Mead, D. L. (1996) 'Determination of available credits and service areas for ESA vernal pool preservation banks', in Carol Witham, C. (ed) *Ecology, Conservation, and Management of Vernal Pool Ecosystems*, from Proceedings of 1996 Conference of the California Native Plant Society, Sacramento, CA, pp274–281

US Fish and Wildlife Service (1988) *Endangered Species Act of 1973 as Amended Through the 100th Congress*, US Department of the Interior, Washington DC

US Fish and Wildlife Service (1998) *Recovery Plan for Upland Species of the San Joaquin Valley, California*, US Department of the Interior, Washington DC

US Fish and Wildlife Service (2002) *Recovery Plan for the California Red-legged Frog (Rana aurora draytonii)*, US Department of the Interior, Washington DC

US Fish and Wildlife Service (2003) *Guidance for the Establishment, Use, and Operation of Conservation Banks*, US Department of the Interior, Washington DC

8

Business Considerations

Craig Denisoff

Introduction

Conservation banking is the art of balancing business and biology. A conservation bank site must have good ecological and business fundamentals in order to succeed. Both the biological habitat and the market for selling credits must be conducive to banking from the very early stages of the project. Understanding the balance between business and biology is important not only to the potential conservation banker but also to the governmental entities that entitle and oversee this new and developing green industry.

It is fairly well accepted that a high quality landscape in a managed conservation bank can provide a number of ecological benefits for species over the traditional on-site or small, postage stamp mitigation (Fox and Nino-Murcia, 2005). It is also well established that conservation banks offer several economic benefits over smaller, individual-species mitigation projects.

First, the permitting and land management of large conservation banks provides economies of scale, by spreading planning, permitting and implementation costs over more acres, thereby reducing the average per unit cost. Second, by permitting one large project, you reduce the time costs of mitigation and can pass along those savings to both public and private sector clients that need mitigation. Third, because the project is typically established in advance of impacts, the higher ratios applied to mitigation projects due to temporal loss of habitat or uncertainty may be reduced, thus lowering overall costs to the credit buyer and providing a competitive advantage over other mitigation alternatives. Finally, and possibly most valuable, a conservation bank provides certainty to a developer who needs species mitigation by providing severance of liability, thus increasing the demand for the product and providing a competitive advantage over other forms of mitigation. When a conservation bank is approved by the agency, the banker takes on the legal and financial responsibility for successful mitigation.

Even with the inherent economic advantages of larger conservation banks, conservation banks are not always a good business venture. For example, while large-scale retail stores offer many cost advantages over smaller retailers, larger stores do not make business sense in areas with a small customer base or specialized needs. In areas where there are other lower cost alternatives to addressing species mitigation, such as other conservation banks or government-subsidized programmes that offer the same service (species mitigation, severance of liability) at a lower cost, a conservation bank may not be financially viable. Thus, a conservation banker, like any business venture, needs to have market conditions that favour the investment and must conduct the same extensive market research and due diligence that any wise business would perform.

Costs of product

Like all production, a conservation bank has a number of cost factors that go into developing the product. The two primary costs can be broken down into materials and service. Since the primary focus of conservation banks is preservation of existing habitat, the primary material cost is the acquisition or control of land. In cases where restoration or enhancement is required, such as building wetlands habitat for a species or removing non-native invasive plants, other material costs may be incurred. The primary service costs associated with a conservation bank are (1) costs of permitting a bank; (2) cost of reports and legal documentation to support the bank; and (3) the financial assurances necessary to ensure long-term sustainability of the bank. Much of the costs of a conservation bank are the biological, financial and legal expertise necessary to document and ensure the long-term protection of the conservation bank site. The final major cost for the majority of conservation banks is the financial assurances related to future site management. This cost is found in the long-term land management endowment accounts, a non-wasting account, that is established to finance the future management and oversight of the site.

For example, the document or legal contract that establishes the conservation bank is typically comprised of the conservation bank agreement that outlines where the site is, what types of species or 'credits' it will yield, the area in which species credits can be sold, the obligations of the banker, land manager and agencies, and then all the supporting materials to ensure the legal and financial compliance of the bank. The production of the bank agreement is often highly technical and legal in nature and can cost upwards of several hundreds of thousands of dollars to develop. In addition, the negotiations and review time by the agencies, especially in areas which do not have a lot of experience in banking or lack the resources to review the banks, can take years to finalize, thereby increasing costs further.

The largest costs often associated with the bank agreements are the legal and financial documentation and compliance monies necessary to ensure the suitability and ecological sustainability of the site. The primary costs associated with documenting the viability of the site are the title reports and environmental site assessments (e.g. Phase I) associated with the land. These reports and documents are necessary to ensure that the sites can be used for their intended habitat purposes and that there are no other uses (e.g. mineral rights, easements, right-of-way) or encumbrances that would diminish their ecological values. In addition to land title, a thorough investigation of the site to ensure that no past environmental damages or toxics issues which could hurt the target species is also required.

Drafting and monitoring the conservation easement associated with a bank also represents a major legal, and often costly, step in building a conservation bank. A conservation easement or other form of deed restriction is required by the regulatory entities to ensure the future land use is assured through compliance monitoring or reporting. This requires the drafting and review of legally binding easement or deed restrictions. The easement typically requires an easement holder and a responsible party to enforce the easement. In many areas, a governmental entity will be the easement holder. However, in recent years many nationwide and local land trusts and conservation organizations have taken on the responsibilities of holding and monitoring the conservation easement. In many cases the easement holder will require monies to provide reports to the governmental or regulatory body on easement compliance and even monies for potential legal defence of the easement if the terms and conditions are violated. Thus, fees for easement holding have risen quite substantially over the years. Costs for outside parties to hold and manage easements in certain parts of the country have actually reached six figure amounts, and, given the uncertainties associated with future legal defences of these easements, these costs should not be considered unreasonable.

Finally, the other important, and generally expensive, document to draft is the long-term management plan for site stewardship. The land management plan often includes the costs to manage the site for basic activities (e.g. fences, signage, trash clean-up, insurance, taxes, etc.), along with costs for more active future management if necessary (e.g. removal of exotic vegetation, water control structures, grazing or burn management, etc.). Given that preservation of species habitat and the future viability of habitat is the cornerstone of conservation banking, this provision gets a lot of attention in conservation banking agreements. In some instances, the costs and responsibilities of managing the land will be assumed by a local, state or federal governmental entity or a local land trust. However, in the majority of conservation banks today, an endowment account is established to pay for the yearly land management of the site.

The endowment account required to fund the long-term land stewardship of the site is one the three major cost items of a conservation bank and in some instances can exceed the costs of the land and the permitting of the bank. These non-wasting endowment accounts are separate financial accounts, typically held and managed by a third party (e.g. government agency or non-profit foundation), in which the interest only, minus the costs of inflation, are used to pay for the annual site management and monitoring. This may seem like a minor expense, but can be quite high. For example, if a site requires an estimated $20,000 per year to manage, then an initial deposit of $400,000 would be necessary at a 5 per cent net interest rate (gross interest minus the Consumer Price Index) for the endowment account.

It should be noted that the party holding the endowment account and the interest rates they obtain impact heavily on the price of production for a bank.

For example, if a governmental body holds the endowment in an interest bearing investment such as money markets, they typically receive a 2–3 per cent net return versus foundations and non-profits that typically hold money in conservative foundation type investments with net industry average rates of 4–6 per cent. These same differences in net percentage rate of returns can result in a doubling of costs for endowment accounts. For example, as described earlier, if $20,000 a year is needed for land management, if it is held by a foundation or non-profit yielding a 5 per cent net rate of return the amount was $400,000, but if it were held by a governmental entity yielding 2.5 per cent net return then the account would need $800,000 to achieve the similar return, twice the amount if managed in a balanced portfolio.

While the average cost of service associated with a conservation bank is impossible to estimate given the variety of costs associated with each project, costs for service can run from the tens-of-thousands of dollars to over a million dollars on projects that require substantial permitting, documentation, easement and endowment fees. Thus, the costs of service and the costs of the materials, which are the land plus any capital improvements necessary, comprise the cost of production for a conservation bank.

Once the cost of production has been determined, it is time to factor in what the demand is for the product and what the market is willing and able to pay.

Market analysis

The market for endangered species mitigation and conservation banks is one of the fastest growing fields in environmental mitigation today. Unlike wetland resources, which are relatively fixed in terms of land coverage, endangered species are related to areas of the country where you have substantial population growth and economic development. Generally speaking, where you have growth, you will see development impacting existing species habitat in a manner that threatens the overall health and sustainability of the species.

Demand for species credits is determined by:

- general market conditions;
- customers;
- regulatory environment;
- competition.

Once the general demand or market for species mitigation is determined, then good financial forecasting and modelling need to be applied to determine the value of the species credits and whether the large financial investments of capital and expenses are justified by the revenues.

General market conditions

Since the credits or habitat values produced at conservation banks are typically sold on a regional basis as determined by the impacted species, it is important to understand the regional economic conditions in the area where the bank operator intends to sell credits. As with all market-based commerce, the strength and diversity of regional economic conditions will dictate the business opportunities. Understanding issues such as projected regional population growth, economic growth, economic base and land use plans is important when it comes to determining the strength of the market.

Population growth: Species impacts, and hence demand for mitigation, are often related to human population growth. In looking at the growth rates from the various census data sets, it is important to analyse the data for real numbers versus just percentage growth and to look for growth in which land will be impacted. For example, population growth in a highly urbanized area may be accommodated by increased housing density versus growth in a suburban or rural area which may result in greater land use.

Economic growth: Economic growth will also impact land use and hence, potentially, mitigation demand. Strong economic growth will stimulate job growth, resulting in a need for more workers and housing. Residential housing along with the infrastructure to support housing and commerce are among the strongest demand drivers for mitigation. Regional forecasts on future growth are an important indicator of future mitigation demand.

Economic base: Since conservation bank credits are typically sold over a number of years, it is important to look to several year demand cycles rather than just one or two year projections. Thus, markets with diverse economic sectors (e.g. service, industry, government) have greater chances of maintaining and sustaining future economic growth than markets with only one or two major business drivers. In addition, markets with a good mix of both public (e.g. government, schools, military) and private sectors are better able to deal with economic fluctuations than areas with only one major industry or employer.

Land use: The area in which growth is projected to occur is also a very important factor in determining the demand for mitigation. Most local governments establish local land-use maps showing where future growth is expected to occur. These maps, when overlaid with habitat or species maps, can determine the type of future impacts that may occur. Areas that are planned as agriculture or open space may be a good location to site conservation banks. It should be noted that agriculture or open space does not automatically translate to good species habitat. For example, golf courses are considered open space but may not make for good habitat. In addition, areas zoned for open space and agriculture often get changed due to development pressures. Thus, there is a benefit of placing a permanent easement and ecological management on these types of lands.

Customers

Understanding the customer base for mitigation credits is another critical factor in determining the future business success of a conservation bank. The two primary customers for mitigation services are private sector developers and public infrastructure organizations.

Private sector: Residential developers represent the largest customer base for mitigation in the US. Residential developers typically require large tracts of lands for their housing projects and are most likely to impact endangered species habitat. Commercial and industrial developers also are users of mitigation, but often their projects are more centralized and thus require less mitigation.

Public sector: The public sector is also a major user of mitigation services. Infrastructure projects such as roads, highways, water projects, sewer lines, transmission lines and other government development programmes comprise a considerable percentage of the overall mitigation market (USC: 23-133, 2001; Denisoff, 2005). Currently, there is a preference in the federal transportation authorization for the use of approved mitigation banks for projects using federal funds. Finally, in times of economic slowdowns, public works projects are often used to increase economic activity and can become a large share of the mitigation market using banks.

Regulatory environment

The other primary, and arguably most important, factor to determine mitigation demand is the level of regulatory enforcement of the existing federal and state laws requiring mitigation for impacts to endangered species and their habitats. How the respective US Fish and Wildlife Service (USFWS) offices implement the Endangered Species Act, along with the state fish and wildlife organizations, will determine the overall demand for species mitigation (USFWS, 2007). While the laws covering endangered species are federal laws, individual regions and states interpret and enforce the laws in different ways. For example, in California impacts to endangered species habitat almost always requires some form of mitigation, whereas in other parts of the US best management practices may be used to address impacts to species and habitat. Or in some cases, little or no mitigation is required.

Political environment: Regions with greater awareness and political sensitivity to environmental issues typically have many regulations and legal protections for endangered species, which lead to strong markets for species and habitat mitigation. The level of local political support for environmental issues is often expressed through the local planning or elected representatives. For example, certain local jurisdictions also have requirements associated with species or habitat, which can add to the demand for mitigation. However, in areas where there is public outcry regarding impacts to a particular species, then many projects may not be allowed to proceed and thus the demand for mitigation may be very low. Interviews with the agency regulators, along with past data on permitting activity can assist bankers in determining the past and future trends in mitigation demand.

Ecological environment: It is widely recognized that in order to have demand for species mitigation, the area needs to have species listed as threatened or endangered. However, the amount of species habitat that is threatened by development as well as the biological status of the species is extremely important to understand. For example, if a listed species is found in an area, but little future development is expected to impact that particular species, then the demand for mitigation may be low. Additionally, there may be little or no demand for species whose populations are extremely rare or are at the brink of extinction, since government policy should not allow impacts to the remaining habitat.

Competition

The existence of a conservation bank does not automatically require that a project use a bank for its mitigation needs. In fact, mitigation for species may be satisfied by a number of different methods, such as permittee responsible mitigation (e.g. project proponents implementing their own mitigation projects) or regional habitat conservation plans or 'in-lieu fee' programmes (e.g. monies in-lieu of the actual mitigation).

Understanding the extent and prices related to other forms of mitigation will dictate the amount of demand that an individual conservation bank site can expect to receive.

Permittee responsible mitigation: Given that no entity is required to use an existing conservation bank, alternative forms of mitigation need to be analysed. Given the lack of existing conservation banks nationally, the major method of addressing species mitigation is through permittee responsible mitigation. Permittee responsible mitigation is when the individual project applicants implement the mitigation themselves. Environmental consulting and engineering firms typically do this because they are responsible for assisting the clients with the upfront project permitting and therefore have a competitive advantage over many banks. Understanding the pricing structure and availability of consulting services in an area will help determine the potential demand for mitigation credits. In addition, having good relations with the consulting community as well as a cost-effective product is important in obtaining a share of the species mitigation market.

Self-mitigation: Another method of mitigation is self-mitigation, where typically a large business or government entity performs the mitigation themselves. Many large development companies and government infrastructure entities, such as transportation or water delivery agencies, contain the internal expertise to implement mitigation projects themselves. In addition, having land or the ability to acquire land, many of these larger entities also have permitting, design, construction and follow-up monitoring expertise on staff and thus may have some internal capacity to implement their own mitigation. Thus, self-mitigation by these large entities can reduce the overall demand for conservation bank credits.

In order to effectively compete with entities that have the ability to self-mitigate, conservation bankers need to be able to provide a very cost-effective product and be able to highlight the severance of liability benefits associated with conservation banks. It is important in these situations for conservation bankers to underscore the 'full delivery' nature of the conservation credit. A conservation credit includes not only the habitat area but also all the permitting, design, monitoring and long-term maintenance cost, along with the legal severance of liability. Often when all these costs, along with the uncertainty factor, are added up and adequately reflected in the price of the mitigation alternative, then the conservation credit price is competitively priced against the cost of self-mitigation or outsourcing to other entities due to economies of scale found in the bank cost structure.

Other conservation bankers: Another major factor affecting the demand for conservation bank credits is the amount of credits available from other conservation bankers. As with all businesses, competition from other similar businesses can greatly influence the market. Understanding the amount and timing of available substitute goods in the marketplace is essential in determining whether establishment of a bank is a wise business decision. The location and preferences of the regulators to the other bank locations can also determine if your bank will have a competitive advantage over other conservation banks. For example, regulators or permit requirements often favour mitigation alternatives that are closer geographically to the point of impact over mitigation that is further away. Therefore, if your bank location is closer to areas of future impacts then it may enjoy a competitive advantage over other forms of mitigation.

In addition, the nature of the competition can also play a role in determining how much mitigation business a banker may receive. For example, private bankers sometimes are at a competitive disadvantage to government-run mitigation banks or programmes, and even some private non-profit mitigation alternatives. Agencies' regulators often have a great deal of discretion over how mitigation regulations are satisfied, and if a sister government agency has a mitigation alternative or a popular non-profit group has a mitigation alternative, regulators are sometimes favourably disposed to choose these alternatives over private for-profit conservation banks. Private 'for-profit' bankers often suffer from the age-old perception that people or groups should not benefit from doing the public good (e.g. health care, education, etc.). Unfortunately, this perception has often had an unintended consequence. One of the original goals of conservation and mitigation banking was to provide value to habitat or species landscapes, so that private landowners would have economic incentives to protect rather than diminish those important resources. When the economic demand or the value-added pricing benefits for these ecological lands are diminished, the financial incentive to protect the threatened or endangered species is reduced. Without the financial incentive to protect species habitat some of these lands will be lost to development or more intensive agricultural or mineral extraction activities.

In-lieu fees/regional habitat conservation plans (HCPs): Another alternative for endangered species mitigation sometimes available to the regulated community is in-lieu fees. An 'in-lieu fee' is a payment to a pre-established fund or programme, in lieu of performing the mitigation yourself, and the programme then takes on the responsibility to implement the mitigation. Most in-lieu programmes do not acquire or even identify the mitigation land until enough funds are collected to implement a project, which is often many years after the actual impacts occurred. This type of in-lieu fee mitigation alternatives for endangered species, which are relatively common in wetland mitigation, often occurs when a regional HCP or programme is available. Regional HCPs are often administered by a local government entity or non-profit organization. While these programmes are seen as a proactive method to direct where development and its resulting mitigation should go in a comprehensive manner, they often conflict with and force conservation banks out of the market. Thus, in areas with a regional HCP the conservation banker has to determine whether it can sell its product or be able to compete in a regulated market.

Threats to conservation banking

Like any highly regulated market, there are a number of threats that can affect the success of conservation banking. The majority of these threats come from the same rules, regulations and government agencies that created the market in the first place.

For example, as discussed, regional HCPs and in-lieu fee programmes run by public agencies or environmental non-profit organizations can supplant the need for private conservation banks. Government run or supported regional HCPs often enjoy a competitive advantage over conservation banks because they can: (1) determine the costs for the product and price it at a level that maximizes revenues, rather than being based on the current cost of mitigation; (2) often have discretion over how and where the mitigation will be satisfied, thus controlling costs; and (3) have regulatory discretion over the permitting process and can direct mitigation to in-house programmes. Besides having a number of serious ecological implications, this scenario often removes any potential of private conservation banking opportunities, undermining the incentive to protect private lands.

Similar in-lieu fee programmes found in the wetland mitigation arena have come under criticism by various reports (US General Accounting Office (USGAO), 2001; Environmental Law Institute (ELI), 2002), and there are proposed rules to disallow the use of in-lieu programmes that have lower standards than other forms of mitigation (USGAO, 2001; ELI, 2002; US Army Corps of Engineers, 2005). However, no such similar study or analysis has been done of species-related in-lieu fee or regional HCP programmes. The majority of these programmes are implemented by governmental or environmental organizations which are often allowed to have lower standards than individual mitigation requirements and/or delay the implementation of the required mitigation. To level the playing field, these inequities will need to be addressed by the appropriate regulatory agencies and environmental groups.

Due to the escalating costs of lands in rapidly growing areas, some local government entities are opting to go with a system that requires the private sector to provide the land and any restoration and monitoring that go along with the mitigation to reduce the liability of implementing these programmes. For example, in northern California the highly successful Natomas HCP, after many years of collecting a straight fee for the mitigation, now requires that entities provide the land along with a fee to manage the properties. This ensures that the HCP can provide the required services even in the face of escalating base land costs. These types of programmes do allow the conservation banking model to work; however, it can result in a dampening effect on prices and thus reduce incentives for entities to deploy capital for conservation banks. The reduction in price is due to the focus on the lowest-cost bid and competition among the competing entities, which can be beneficial in terms of pricing, but can also lead to lower ecological values since the value is on price and not on best habitat or quality of product. As astronaut Alan Shepard was quoted as saying regarding space flight 'It's a very sobering feeling to be up in space and realize that one's safety factor was determined by the lowest bidder on a government contract' (www.brainyquote.com/quotes/authors/a/alan_shepard.html).

Another growing phenomenon is the use of local governmental restrictions to reduce or limit the sale of conservation credits. Certain local governmental jurisdictions are concerned that more urbanized areas will use rural lands to meet their mitigation needs, thus limiting the ability of these rural areas to grow or mitigate their own development decisions at a future date. While most conservation banks are situated in areas zoned for open space or other compatible land uses (e.g. recreation, agriculture, etc.), local governments are expanding their typical use of land use controls to limit sales of credits to areas of their jurisdiction. Recent court challenges (*Calmat Co. vs The City of Colton*) are underway to address this issue, but this again has the ability to threaten the future market for conservation bank credits.

Finally, delisting of a species due to recovery or loss of the species will also remove the need and market for conservation banks. While it is everyone's desire to see species recovered to the point of delisting, it will result in the loss of demand for mitigation associated with the species and must be considered in the evaluation of whether or not to establish a new conservation bank for a species that may be delisted in the near future.

Pricing and marketing credits

Understanding the pricing and marketing of conservation bank credits is an important factor in developing a sustainable banking business. Similar to other businesses, understanding who the clients are and the service or product you can provide them is paramount to success. Unfortunately, given the 'green nature' of this industry, many individuals and groups neglect to focus on the underlying business fundamentals associated with the conservation credits and sometimes run the risk of failure.

Pricing the conservation credit

Given that consumers of mitigation have many choices in fulfilling their mitigation requirements, prices for banking credits are typically set at a market rate based on costs of production and available alternatives. There is a commonly held misconception that approved conservation bankers can charge exorbitant prices for their credits, but as described earlier there are a number of alternatives available to keep costs down. Due to the additional benefits that an approved conservation bank can offer, such as reduced permitting costs, timely implementation and severance of liability, banks can often charge a fee that may be slightly higher than the alternative costs, but typically only by a small margin.

Bankers typically set their rates at levels that are similar to or slightly above the costs of individuals to implement the projects themselves or hire contractors to perform the work. However, price elasticity also plays a factor in mitigation demand. Given that the mitigation market is subject to the interpretation and implementation of laws, if prices for mitigation become too costly to the mitigation community, potential credit buyers may undertake efforts to revise the laws or develop lower cost alternatives (e.g. in-lieu fees, regional HCPs). Bankers must take into account whether or not the eventual cost will be too pricey for the market to bear, resulting in the deferral of projects or efforts to revise the implementation of the regulations.

Marketing the product

The market of habitat values or 'credits' differs from general sales and marketing of most consumer products. This is a service industry requiring a high level of expertise and knowledge of the regulatory field and the specific species requirements. In addition, issues of integrity of the product you're selling and the limited nature of the business or organizations that require mitigation dictates the approach to marketing.

Services vs product approach: The regulatory process and requirements are very complicated and not an easy process to get through. Most customers who require mitigation come from fields outside of the environmental field and are not intimately familiar with ecological requirements of species and the requirements associated with species mitigation. Often, they require assistance to determine whether they can buy a mitigation credit or must avoid any impacts. In addition, agency regulators in areas with high development activities have little time or lack the resources to fully assist the permit applicants with all the regulatory process. So while an approved conservation credit is an individual unit or product for sale on the open market, it takes a highly skilled and knowledgeable person to truly market and sell the credit. Even the processing of the sale requires government approval and tracking to ensure mitigation compliance. Therefore, bank sales and marketing individuals trained in environmental regulatory permitting and compliance are essential in being able to assist clients with satisfying their species mitigation needs.

Full delivery product: As described earlier, a conservation credit includes land, legal protections, permitting, design, restoration, monitoring, long-term stewardship and transfer of legal responsibility to a third party, all wrapped up in one unit. Most individuals implementing their own mitigation have to pay for each of these aspects of mitigation separately, often at higher unit cost for each item. However, many consumers of mitigation credits don't know or understand all the costs or benefits that go into a conservation credit and may consider this full delivery product more expensive than the alternative mitigation products which may or may not describe all the future costs associated with mitigation. It is important for bankers to highlight all the products and services that are rolled into a conservation credit to their customer to fully emphasize the benefit of their product.

Product certainty: As described earlier, under the current system, conservation bank credits are not freely transferred among individuals but rather are dictated by the agencies requiring mitigation. This transaction, which requires the buyer and seller to have an agreed upon price and sales contract and the final approval of the governmental agency, is a 'triangular transaction'. While the price and contractual relationship between the buyer and seller is a market transaction, the ability to approve the transaction and ensure that the buyer has legally satisfied their mitigation obligation resides with the governmental agency. Thus, it takes three parties to consummate the sale. It is very important for the conservation banker to keep impeccable files of every transaction and to ensure that the governmental agencies have accurate and up-to-date files to avoid transaction uncertainties.

Approaches to marketing and sales

Since not all individuals require a conservation credit in the course of their daily activities, the conservation banker needs to concentrate marketing efforts on those entities that require mitigation in the public and private sector community.

As described in the section on Customers, the major users of mitigation are private sector home builders, commercial developers and industry and public infrastructure agencies such as road builders, water purveyors and utilities. It is important to target these industries and their associated groups such as consultants and engineering firms, along with land use attorneys and other businesses that assist the core customers in meeting their mitigation needs.

Focused marketing such as trade conferences, specialized mailings and presentations that target these groups is necessary to let them know about the products and services that a conservation bank provides. In addition, knowledgeable and experienced sales staff that understand the permitting process can help meet the client's requirements.

Unfortunately, there are many bankers who enter the market from other industries and attempt to inappropriately apply sales techniques. For example, the traditional 'glad handing' and networking approach has little or no value unless the person can actually help the client with their mitigation problems and understands how to get through the process. Clients for mitigation are extremely busy and their time is valuable, thus talking heads do nothing but waste their time and it reflects poorly on the product.

Traditional marketing tools like billboards or telemarketing that attempt to reach a broad audience are highly inefficient and can even cheapen the perception of the product with the clients and the governmental agencies that are often peripherally involved in where credit sales may go. Marketing and sales that focus on product integrity and servicing the client mitigation needs will ensure long-term success in the industry.

Future opportunities

While the overall market for conservation banks is still very small and limited to only a few areas of the US, there may be opportunities to grow the market for this green product.

Conservation market for credits: While the economic axiom Say's Law states that supply creates demand, it is widely accepted that conservation credits are controlled by the regulatory enforcement of mitigation. What if, though, conservation credits could be marketed to those groups or individuals who simply value conservation?

Currently, there are no restrictions on selling conservation credits to those who want to simply buy credits for the sake of protecting or restoring the environment. Given that most banking credits are full-service environmental units, in the sense that they provide not only the protected land, but monitoring and long-term land stewardship built into the final good, buying credits may be an effective strategy for achieving land conservation goals. Most non-profits or government agencies currently acquire only a portion of the environmental services offered in a credit. Most land trusts or governmental organizations will acquire the land, but then turn the land over to someone else to manage, or will restore existing public lands. Fewer will actually create an endowment fund for the long-term stewardship of the lands, and this activity is usually subject to public budgets and financial fluctuations that impact governmental financing. However, a mitigation or conservation credit typically contains all of these features in one. Once the benefits of these full-service credits are more fully understood by decision makers and the public, opportunities to market credits on the open market may become a more viable option. As described earlier, the costs associated with most conservation credits reflect the actual market costs of providing for full-service ecological protection. Potentially, expanding existing conservation credits to governmental grant programmes may open the door to this market.

Growing need for species credits: While no one hopes to see greater loss of habitat and more species listed, the increasing loss of species habitat seems inevitable as the population continues to expand. Population in the US just passed the 300 million mark and is expected to reach 400 million by 2043; in states and regions where growth is likely to occur, the use of high quality conservation banks will probably increase (US Department of Commerce, 2006).

International market: It is surprising that, with all the supposed attention paid to the environment by developed and newly developing countries, conservation banks are not found in other areas. In some older developed regions there may not be many species left to impact, such as is the case in parts of the north-eastern US, or growth controls have limited impacts to natural resources. But in other natural resource rich countries, such as Australia, Canada, Chile, China, Costa Rica, Japan, Mexico, New Zealand, South Africa and Uruguay, the opportunities to enact laws to protect species could lead to opportunities for conservation banking. See Chapter 13 and Chapter 14 for more on international developments.

Finances associated with conservation banking

While many individuals, businesses and governments have tried to estimate both the general cost structure and profits associated with mitigation and conservation banking, there is no 'one size fits all' approach. All projects have different materials costs and service costs (e.g. permitting, legal and financial assurances) associated with the conservation bank. Thus, it is often more appropriate to use a return on investment model, similar to one a developer would use on a project, to determine the viability of a project.

As every economist or developer recognizes, a simple calculation of a per cent return over the cost of production is not useful in the case of a future return on investments where risk is involved. For example, a simple mathematical calculation shows that if a banker invested $5 million dollars in a conservation bank and received $10 million dollars in net pre-tax profit this would result in a 100 per cent return on investment. However, if it took a banker five years to sell those credits the annual return on investment would only be 20 per cent, and if it took a total of ten years the return on investment would be 10 per cent. If you assume that the cost of capital (the rate a bank charges for the use or their money) for the conservation banker is 10 per cent, then a bank that takes ten years to sell provides no return over the cost of money. When you consider all the market risks of selling credits over a ten-year period, the banker would have been better off investing his or her money in a balanced investment portfolio.

While this example provides an important constraint to the prospective conservation banker, it also is an example to the government regulators. The most standard example bankers see today is when regulators attempt to reduce the amount of credits on a yearly basis or want to restrict the size of sales at a conservation bank to only small impacts (e.g. less that an acre of species impact). These types of regulations, while possibly having good biological rationale if applied to all forms of mitigation, can and do often result in much lower returns and the creation of fewer banks.

The few general characterizations that can be made concerning the potential costs and profits of mitigation banks suggest that unequal requirements by government agencies for different types of mitigation have substantial impacts on conservation bankers. Given the cost of money, the faster a banker can sell their credits the more profitable they are going to be.

Foundations of a successful conservation banking business

There is no one characteristic or feature that makes one conservation bank or banker successful, rather it is a number of different and diverse skill sets that set certain bankers apart from others. Thus, the foundations for developing a successful conservation banking business are:

- capital;
- multi-disciplinary staff;
- internal expertise;
- local knowledge;
- honesty and integrity.

Capital

Conservation banking is relatively capital intensive, often requiring the purchase of property in rapidly growing areas with rising land costs, substantial costs for permitting and design and construction (if necessary) and a large endowment account for future land management. In addition, given that most of the credits associated with conservation banks are sold over a number of years rather than in one or two years, the carrying costs of money can be substantial. The term 'patient money' is often used to describe the type of investment dollars necessary to support the establishment of a conservation or mitigation bank. Rates of return often associated with venture capital or hedge funds are difficult to support given the localized markets associated with species or wetlands banks and the level of competition.

Multi-disciplinary staff

The permitting of a conservation bank takes a number of different areas of specialized expertise. Fields as diverse as biology, ecology, real estate, legal, geographic information systems, regulatory planning, government process, marketing and sales, and financial accounting are often necessary in the establishment of a conservation bank. Understanding all the different factors or acquiring the necessary expertise is fundamental to banking.

Internal expertise

While the fundamental aspects of how to permit a conservation bank and the bank agreements are relatively standardized in the parts of the country that have conservation banks, there is still a great deal of variation in the length of time it takes to permit banks and the type of requirements that go into individual banks. The length of time it takes to permit a bank, often from six months to a couple of years (some conservation banks have taken up to four years to permit), and the differing amounts of expertise can result in substantial costs to get the product on line. Thus, if you have to outsource a substantial portion of the work it can cost a lot of money and dilute any potential profits or cause you to raise your costs. In some instances, the increased standards and requirements price out the smaller landowner banks that cannot afford the upfront costs associated with bank development.

Local knowledge

Like politics, all banks are local. Conservation bank credits are sold for a particular species that in most instances is found in a highly localized area. Service areas are typically restricted to one or two counties. In some instances, the service area or area in which credits can be sold is limited to a city or portion of a county. Knowledge and involvement with the local regulatory community, decision makers and customer base is vital in establishing and marketing your credits. Bankers or mitigation providers from outside the area are at a competitive disadvantage when it comes to contracting local groups to provide business services.

Honesty and integrity

Finally, the one characteristic that sets bankers apart and determines who will be around over the long term versus those who are 'here today, gone tomorrow' is honesty and integrity. A banker may make a large one-time sale or even completely sell out a bank, but if they do not provide honest service to the client, the client will choose to take their future business elsewhere. In addition, if the credits sold do not meet the regulatory expectations of the agencies or the product is oversold, then the agencies will be less likely to permit a future bank.

In the end, the key to a good project is similar to most other business ventures and requires the following components to be successful:

- Capital: land or investment monies.
- Expertise: ability to permit the project.
- Opportunity: client plus regulatory environment plus good landscape.

Capital + Expertise + Opportunity = Success

Summary

Banking is truly the intersection between business and biology. Economic factors and business factors have to be weighed as carefully as the ecological and biological factors before establishing a conservation bank. The actions of the regulatory and government agencies can have major impacts on the business of banking and need to be monitored closely. In addition, governmental bodies need to pay close attention to the policies and regulations on mitigation and should understand the economic impacts of their requirements on all forms of mitigation if they want to maintain conservation banking as an ecological alternative.

As the population of the US continues to grow, so will impacts to endangered species and their habitats. Therefore, the need for high-quality and cost-effective species mitigation such as conservation banks will continue to expand. The expected growth in the use and demand for conservation banks will lead to many more landowners and groups, such as non-profits, land trusts and private commercial concerns, providing these services. However, it is important for all parties engaged in the establishment and management of conservation banks to recognize that it is a complicated business that requires both a solid foundation in biology as well as refined business skills to succeed.

References

Calmat Co. vs The City of Colton, Superior Court, San Bernadino County, Case Number SCVSS135476

Denisoff, C. (2005) 'Banking and transportation projects: Merging ecological protection and economic growth', *National Wetlands Newsletter*, September–October

Environmental Law Institute (2002) *Banks and Fees: The Status of Off-Site Wetland Mitigation in the United States*, ELI, Washington DC

Fox, J. and Nino-Murcia, A. (2005) 'Status of species conservation banking in the United States', *Conservation Biology*, Society for Conservation Biology, pp996–1007

US Army Corps of Engineers, 'Compensatory mitigation', Report (unpublished) prepared in 2005 for the Directorate of Civil Works

USC (2001) *Transport Equity Act for the 21st Century*, 23 USCA, Title 133, Thomson West, Eagan, MN

US Department of Commerce, Bureau of Economic Analysis (2006) 'Gross state product data table', *Regional Economic Accounts,* October

US Fish and Wildlife Service (2007) *Consultation with Federal Agencies: Section 7 of the Endangered Species Act*, www.fws.us.gov/endangered/consultations/s7hndbk.htm

US General Accounting Office (2001) *Wetlands Protection: Assessments Needed to Determine Effectiveness of In-Lieu Fee Mitigation*, GAO-01-325

9

Financial Considerations

Sherry Teresa

Introduction

Conservation banking is an innovative and creative tool to assist in global conservation efforts to protect threatened and endangered species and their habitats in perpetuity. In order for conservation banks to successfully and perpetually compensate for losses of species, there needs to be concise and clear-cut agreement between the banker, permitting agencies and long-term steward regarding the development, financing, credit release schedule, operation, the long-term management and the funding source of the bank. Conducting a thorough due diligence is the first step for a banker and bank steward to ensure the maximum opportunity for success. Conservation banks, by their very nature, are lands set aside for the permanent protection of native species and their habitats. The intergenerational protection of land can most effectively be accomplished through the use of conservation easements.

But merely setting the land aside is not sufficient to protect its resources. Without active long-term stewardship, those conservation values can be rapidly lost to invasive species, trespass and unauthorized or destructive uses, urban encroachment, habitat conversion or a myriad of other threats. So how does one identify and develop a long-term stewardship programme to protect those resources? And how does one determine what that stewardship programme will cost now, ten years from now, or 100 years from now?

Estimating land management costs in perpetuity is challenging. Virtually all costs and revenues associated with conservation banking are dependent on the location of the lands being conserved. There are no typical costs. There are no rules of thumb. There are no averages. There are no set rates. Every input is driven by the marketplace and the required tasks. The Property Analysis Record (PAR©), developed by the Center for Natural Lands Management (CNLM), is one tool that has been widely used to estimate tasks and costs associated with long-term stewardship, project cash flow, identify various costs for phased banks and determine an endowment amount. These items will be explored in more depth in this chapter.

An endowment is a funding source that provides long-term assurances and income for stewardship activities. However, without sound financial assumptions used in establishing endowments, such as keeping the endowment current with inflation, they may become 'wasting endowments' and are not a perpetual funding source. To ensure proper use and reporting of the funds, they should be managed, invested and accounted for using standard rules and accounting principles, with annual independent audits and transparent reporting formats.

Is it fair for bankers to transfer all risks related to future ownership and management of conservation banks to stewards? Can any amount of due diligence uncover or anticipate all conditions or contingencies? Is it possible to adjust initial and capital costs, ongoing costs or endowments to compensate for this transfer of risk? These and other such questions will be discussed in this chapter. This chapter will also describe:

- how to conduct a due diligence and estimate the long-term costs associated with stewardship of lands dedicated for conservation banking;
- the types and uses of instruments to fund those costs;
- how to anticipate and manage financial risks associated with conservation lands.

This chapter is written from the perspective and experience of CNLM, a California-based non-profit organization that owns and/or manages over 17 conservation and mitigation banks. While it is not possible for a single chapter to address the multitude of investment, regulatory and stewardship issues that arise for every type of US and non-US organization, the basic investment principles enunciated here are generally applicable worldwide. Non-US organizations should consult local authorities and advisers regarding the implementation of their regulatory requirements relating to investment and endowment programmes.

Banking agreements

A well-conceived and thoughtfully considered conservation bank has a bank agreement signed by the bank sponsor, agency representatives and the long-term steward. The conservation banking agreement should define precisely how many credits will exist when success criteria are met, and where and when credits may be used by purchasers. From the wildlife agencies' point of view, bankers frequently keep inadequate records of credit sales and frequently claim their land has more conservation values, and therefore they should be granted more saleable credits than was initially proposed. There is nothing wrong with such 'multi-species conservation banks' and many banks have successfully added species and thus created additional credits. The missing links are an update to the credit-sale registry and provisions for additional management funding for new tasks added to the steward's scope of work.

Typical documents found in conservation banking projects are:

- conservation banking agreements;
- documents related to acquisition and transfer of land – such as purchase and sale agreements, title reports and title insurance, escrow instructions and deeds;
- documents assuring perpetual protection of resources – such as conservation easements;
- bids and contracts for construction and monitoring success criteria;
- management and funding agreements for long-term stewardship;
- habitat management plan;
- credit sale registry, approval process and master escrow.

In order for a conservation programme to be most effective the documents involved in the transfer of rights to land, the grant of rights to sell credits, establishment of conservation goals – including the means and methods of preserving conservation values – and funding should be integrated. For many reasons, most of which relate to the complexity of gaining approvals and the length of time between application for approval and commencement of credit sales, this is rarely done. Problems generally occur in the sale of credits and transfer of funds required for long-term management.

From the bankers' perspective, wildlife agencies grant approval to construct conservation banks, but then delay approvals or refuse to allow the credits to be used to mitigate other projects. From the stewards' perspective, funding sources for long-term management need to be established 'up front', and augmented when additional species, and additional work, are added to a conservation bank. Many stewards would like to cooperate with bankers and defer funding for management until such time as credits can be sold. However, in more cases than not, such deferral has weakened the financial capacity of the steward and jeopardized the protection of the resources. Depending on the nature of funding for long-term management, deferred payments can jeopardize a steward's ability to pay future costs.

Although there are many players in a conservation bank, rarely are they all present at the negotiating table from day one. Rarely do all sign each of the important documents. Rarely is there a ready means to enforce provisions of the agreements. In any case in which payments are deferred, one party, typically the steward, is acting like a lender to another party. Commercial lenders have policies for documenting and securing the lending and collection of money. Most stewards do not.

Similarly, wildlife agencies, in approving the creation and sale of conservation credits, are allowing the banker to create something of value (the credits), without full security for performance. Later, this chapter will describe the use of payment and performance bonds for long-term stewardship of conservation banks. In addition, permitting agencies should consider holding a pledge of the credits, secured by a filing or lien – depending in which state the credits are located – which is released, upon sale, only when all conditions precedent to such sale have occurred. Those conditions include acquisition of land and assurance of perpetual conservation; completion (in phases if so permitted) of restoration, enhancement or construction; sign-off on success criteria; provision for long-term provision for and funding of long-term management.

Due diligence

Due diligence is the research and assembly of all relevant information that allows assessment of the feasibility of a project and helps to ensure its long-term financial stability. The time involved in due diligence is likely to be lengthy. These capital costs include:

- identification of land and biological consulting costs to determine if threatened or endangered species or their habitat are present;
- evaluation of title and exceptions;
- environmental site assessments, phase 1 and phase 2 investigations related to hazardous materials and related liabilities;
- cost of the option to purchase land;
- costs associated with the option, such as taxes and insurance;
- engineering and construction studies related to cost to restore or enhance habitat;
- legal counsel and input on the cost and time to get through the regulatory and permitting process;
- fees of a steward in preparing a long-term cost estimate;
- cost of hiring consultants to perform an economic feasibility analysis, research the market for current and future credits, and analyse costs and revenues.

If due diligence reveals that a project is feasible, the regulatory process can commence; upon issuance of a permit, the option to purchase land can be exercised, and restoration, enhancement and/or long-term stewardship can commence.

Acquiring and maintaining rights to the land

Conservation banks must be protected in perpetuity through title restrictions. How this is done can affect the long-term stewardship costs of a bank. The bank sponsor can own the land outright in fee simple title, or may retain a conservation easement over the bank (if they are a qualified entity under the tax code). In several cases, bank sponsors have relinquished fee title or any holding on the land to a non-profit or public agency but retained the right to sell the credits. Credits may be severed from the title similar to mineral or water rights.

Although fee simple ownership has the fewest legal restrictions on use and transfer, such ownership is subject to certain exceptions, most notably the right of the public to levy and collect taxes; the right of the public to acquire the land (eminent domain); the right of the public to the land if its owners have died with no heirs (escheat); the right of the public to impose restrictions (zoning and other restrictions on use – also public health and safety requirements); and limitations imposed by the current owner or prior owners (conveyances of mineral or water rights, deed restrictions, easements, servitudes, licenses, leases, mortgages or deeds of trust securing the performance of obligations). All of these can have impacts on the long-term costs of maintaining a bank. In many states, non-profits are not exempt from property taxes, special districts and other fees (i.e. water, fire, vector control, etc.). Any thorough due diligence would identity these fees and allow for their payment in an annual stewardship programme.

Fee simple title to land may be held by any legal entity or combination of entities – alone or together (for example, an individual or individuals otherwise lawfully eligible to own property, general or limited partnerships, trusts, corporations, limited liability companies (LLC), unincorporated associations, etc. While there are many forms of ownership, the topic is beyond the scope of this chapter. Because of the nature of LLCs or private ownership, an irrevocable conservation easement should be placed over every conservation bank, regardless of the ownership. For example, land may be held by the public, but a conservation easement from the public to a third party private entity concerned with conservation may provide stronger assurances of timely monitoring and management. Conservation easements may become invalid if not superior to any loans or liens on the property's title. Agencies permitting conservation banks should require the granting of a conservation easement before credits may be released and ensure that the easement is not subordinated to any title encumbrances that would jeopardize it.

One should be careful when allowing the use of deed restrictions or covenants to be used to protect banks. Covenants and deed restrictions contain many of the same elements as conservation easements but typically lack monitoring or third party enforcement by a grantee or third party beneficiary. Covenants and deed restrictions may be revocable. In some jurisdictions, they are not interpreted strictly or enforced by courts to the same degree as conservation easements. In constructing a conservation programme, all of the options available for the acquisition of rights to protect land can and should be considered. Consultation with a land-use attorney, real estate and/or estate planning attorney and other professionals can be useful for evaluating the alternatives.

Third party beneficiaries and enforcement

In most conservation banks, if a public agency is not the conservation-landowner or responsible for long-term management, one is named as a third party beneficiary of a conservation easement and the funds set aside to maintain, monitor and manage the land. Enforcement rights are granted to the holders of conservation easements, or to some third party, where there is no conservation easement, and land is owned by a banker or steward. Yet in most cases it is preferable that legal enforcement be delegated to and accepted by a public entity or agency that can investigate, prosecute and collect damages.

If agreements are cross defaulted, as described above, and if a banker is in default, the permitting agencies will have an easier time stopping credits from being sold or used until the conservation values are no longer in jeopardy. And if a steward is in default (as was the case with The Environmental Trust in California [Teresa, 2006]), the agency can more quickly assume or transfer management and monitoring, along with funds set aside for those tasks.

It is highly recommended that all tasks related to construction, restoration or enhancement be completed prior to assumption of long-term stewardship, and all monies necessary for such long-term management be assured or collected up front.

Property taxes and how they relate to stewardship funding

In most jurisdictions, a title report shows the assessed value of the property for property tax purposes, delinquent property taxes owed, and the amount of property taxes currently payable and their due date. A title report may also show assessments and other liens of special districts that may affect the land. However, in California for example, this is no longer the case and a current tax bill must be reviewed for a list of special districts and fees. If a lien or notice of lien has been recorded against land, but no tax is shown on the title report, or the report states that such assessments and liens may be of record but are not disclosed, a party with a prospective interest in the land should do further research to determine whether the land is subject to such taxes.

Property taxes are a complex subject and vary from state to state, but a few insights are important. First, land and improvements tied to the land are subject to property taxes based on value. Such taxes are known as *ad valorem* taxes – based on value. Land and improvements are also subject to assessments. These taxes are frequently collected with the *ad valorem* tax, and pay for improvements to land, or improvements near land – such as roads, drainage, flood control, etc. Finally, special districts or community service districts also collect taxes. These districts provide a myriad of services such as landscape maintenance, street lights, police, fire, schools, parks and recreation.

Many states do not have mechanisms to relieve non-profits from property taxes and most land trusts will only hold conservation easements, not fee title. Assessments and special district taxes may still be due and payable. Even when conservation land is deeded to a non-profit or a governmental agency, some assessments and special district taxes must still be paid. Qualified non-profit entities concerned with conservation are granted property tax exemptions by some taxing authorities. However, this still may not relieve the land of assessment and special district taxes. Assessments can be paid off, typically by a banker, prior to transfer of long-term management to a steward. Some special district taxes cannot be prepaid. In this case, some provision must be made for their payment.

Finally, the parties to a conservation bank want to be assured that taxes and assessments are paid as they come due. The government can collect unpaid taxes by selling real property at tax-sale or tax-foreclosure. In some cases, that sale or foreclosure can eliminate the goals and protections of conservation banking agreements and conservation easements. The successful buyer at sale may own the land free and clear of such restrictions. The conservation values may be in jeopardy without a recorded conservation easement on the property. This is an exceedingly complex topic requiring all parties to seek counsel experienced in local land use law and taxation, but it should not be overlooked. As with all creditors whose rights are secured by real property, the rights of taxing authorities can defeat an otherwise well-planned conservation programme.

Are conservation credits taxed?

Typically if owned by a taxpaying entity, the gain on the sale of conservation credits (revenue minus cost) is taxable. Conservation credits may also be taxed the same way mineral rights are taxed. Indeed, conservation credits, since they are used up over time, are not unlike minerals, oil and gas. In some jurisdictions, unsold conservation credits may be subject to some kind of personal property tax.

In one instance in California, a conservation banker was losing his property in a tax sale but the credits were severed and viewed as personal property. The bank had a recorded conservation easement which would have survived the tax sale, so the new owner was obligated to comply with the easement terms and conditions. The bank sponsor was able to repay his liabilities and prevent the sale. However, it did bring to light interesting questions as to how bank credits are viewed in different situations.

Estimating and funding long-term stewardship costs

The cost of preserving, protecting, restoring or enhancing land is based on the current and anticipated future market price of labour and materials. Since every parcel of land is unique, based on its specific location, these costs are going to vary from place to place – based on local labour and materials market conditions, and based on the ease or difficulty of accessing, working on and managing the specific parcel in question. Those charged with stewardship, the long-term monitoring and management of conservation lands, will incur costs – many based on strictly local considerations:

- labour and wage rates and costs of employee benefits;
- cost of materials and equipment;
- overhead costs such as office occupancy costs, insurance and local taxes.

Inputs not driven by local considerations are interest rates and earnings on investments. These rates tend to be a function of national and global economics. They are key inputs into a steward's long-term calculations, and they are key inputs into calculating opportunity costs.

Assurances that success criteria are met

If the land must be restored, or if the conservation bank involves enhancement in order to generate enough credits and revenue to pay costs, it is highly likely that regulatory or permitting agencies will require some assurance that the work will be completed and all of the success criteria met.

Payment and performance bonds

These assurances or guarantees are typically in the form of performance and payment on completion bonds issued by insurance companies that commonly deal with construction projects. The banker pays a premium to the issuer for its bond.

The legal provisions of such bonds are complicated and legal counsel should be sought for further clarification, but all parties – banker, regulatory agencies, contractors and stewards should be aware that the practicality of such bonds is based on:

1 the financial strength of the bonding or insurance company;
2 the language of the bond or policy as to how and when claims are to be made and paid;
3 requirements for notice;
4 requirements that certain procedures be followed with respect to documenting and paying for the construction work itself;
5 expiration dates;
6 the speed with which action is taken after a default in the project agreements for which the bonds were posted.

Letters of credit

In lieu of bonds, a banker may deposit cash into an escrow account or obtain a letter of credit. All such sureties, indemnities and guaranties, as these kinds of financial instruments are called, have risks and benefits. In most cases, a letter of credit from an FDIC-insured, federal or state chartered banking institution, confirmed by a similar bank of unimpeachable financial strength and integrity, is the instrument least likely to be challenged when drawn, and least likely to be subject to the scope of the insolvency or bankruptcy of any party.

However, letters of credit themselves have terms which must be strictly adhered to – including expiration dates and the timing, means and methods of applying for payment. Letters of credit are generally more difficult to obtain and more costly for bankers than are bonds. They generally have one-year expiration dates, which require those holding them to request their renewal. Hence, if they are not renewed, that assurance can be lost. Legal counsel should always be used when obtaining or using financial assurances such as the pledge of cash or property, payment and performance bonds or letters of credit.

Flexible use of financial guaranties

Financial guaranties can generally be tailored to the precise needs of the parties. This might include:

- Timing amount and delivery of the assurances to coincide with actual construction schedules and phasing of the project.
- In certain cases, allowing credits to be sold in advance of full success – dependent upon adequate progress and financial security – for example, a guaranty in the amount that would be required to purchase substitute credits from another bank.
- Guaranty amounts that decline over time, as work is completed and success criteria met.
- Pledge of credits in lieu of other collateral or financial guarantees. This might occur at a phase of completion where some credits can be released for sale, but some work remains to be completed.

Success criteria dependent upon biological considerations are covered in other chapters. But generally, a resources agency or a biological consultant independent of the owner and the restoration/enhancement contractor is charged with evaluating construction during a three to five year 'monitoring and maintenance period' (M&M period) following completion of basic restoration/enhancement work. The agency or consultant and any long-term manager or steward should coordinate with the consultants their tasks and reports to avoid overlap and duplicative costs. Credits are generally not available for sale, nor may they be used to mitigate take at other locations until certain success criteria for the establishment of a functioning ecosystem are met.

Stewardship: Determining long-term habitat management tasks and costs

Long-term management of conservation lands often begins before success criteria are met, and is never ending. The steward – as the long-term manager of conservation land is known – is charged with assuring that the conservation values are preserved and protected, and that restorations and enhancements accomplish their goals and objectives.

In most conservation banks, the steward is a public agency or a non-profit entity organized for the purposes of conservation. The steward should have considerable land management, financial and administrative management experience, with a staff or access to consultants skilled at biology, ecology, natural resource management, range, farm or timber management, finance, accounting, investment management and insurance, management and administration, and financial and biological reporting. The financial management is as critical as the biological management to ensuring perpetual stewardship and the bank's conservation goals.

Long-term stewardship

It is possible to reduce the long-term costs of conservation land management to a single value – a financial calculation, in today's dollars – which is the present value of all future costs associated with a project, discounted at an earnings rate that takes into consideration achievable risk-adjusted economic returns on liquid investments, over time, and the impact of inflation on the purchasing power of those economic returns.

The present cost of long-term stewardship is arrived at as follows:

1 Compile a list of management tasks required to be performed now and in the future.
2 Calculate the current cost of those management tasks.
3 Construct a financial spreadsheet, by timing of occurrence (weekly, monthly, annually, every 10, 15, 20 years?), adjusting the costs over time to allow for anticipated inflation and discounting at a rate equal to what can be reasonably earned on liquid investments.

This would require a very detailed spreadsheet and, since long-term management has no termination date, entries out into perpetuity.

Property Analysis Record

CNLM (www.cnlm.org) developed an alternative model that is far more accessible to most stewards. It is called Property Analysis Record, or 'PAR' for short. The Center offers PAR software under a licence, with appropriate training. PAR has been adopted for use by many wildlife agencies, stewards, non-profits concerned with conservation, biological consultants and contractors throughout the US. It is used as a due diligence model in addition to tasking and costing habitat stewardship. The PAR facilitates communication by translating biological and protection requirements into the common language of currency.

PAR works by first evaluating the long-term needs of the project by comparing a set of management possibilities against the actual condition and circumstances of the project. This results in queries such as: Will the water source need to be supplemented? Can it be contaminated by chemicals or invasive exotic plant seeds? And is there sufficient upland preserved? What are the neighbouring uses? And will they cause excessive trespass, damage, erosion or dumping? These are typical questions to be asked in the due diligence process. Second, PAR asks what can be done in the design or documentation of the project to reduce these threats. The analysis becomes the incentive for improved design because its upfront costs can be directly compared with the long-term costs of stewardship with a lesser design.

Understanding this relationship between upfront efforts and long-term requirements is an important result of utilizing the PAR tool. Third, the PAR asks what tasks are necessary to offset the remaining threats to the project over time. The advantage to the PAR in this step is that it directs the user to evaluate these tasks in terms of their size, unit cost and timing. With these assumptions, the PAR prepares a task-based budget suitable for presentation as an annual cost, an average annual perpetual cost, and as an endowment.

By using these task-based budgets, mitigation providers, stewardship organizations and regulatory agencies are fully informed about the tasks and assumptions making up the cost of permanent stewardship. The Center has found that full transparency in developing and presenting these budgets has been imperative in achieving mitigation sites with fully dedicated endowments. Mitigation providers are far more willing to contribute to an endowment when they can clearly see for themselves that the budget is credible and sensible. The fact that nearly all mitigation projects where the PAR has been in use for several years are backed by stewardship endowments or other funding is evidence of its success.

Implicit in this project is the ability to forge partnerships and to replicate the process. The PAR provides to all parties an objective frame of reference for negotiations, with the ultimate benefit to the resource in both the short (in cost-effective site design), and long (cost-effective management expenditures) terms. The PAR model encourages stewards to break down their management obligations into categories with discrete tasks. These tasks are evaluated at two discrete points in time.

Initial period management – initial and capital (I&C)

Tasks are determined and costs established for the initial period – at the time habitat management operations begin – generally the first one, three or five years of operation. This may coincide with a banker's M&M period, overlap somewhat, or follow sign-off on success criteria. The time-frame for the I&C period will depend on each specific case and the difficulty and complexity of preservation, restoration, enhancement and management. It may also depend on the bankers' land acquisition plan, construction or enhancement phasing and speed of credit sales.

In the PAR model the tasks completed and costs incurred during the initial period are described as 'initial and capital costs'. The PAR model treats them as 'one time' costs even though some tasks may be performed and some costs incurred more frequently than once. These upfront costs are not capitalized into the endowment. CNLM uses a figure of four to five years I&C on all labour costs on PARs using the endowment funding scenario. This allows for any adjustments or downturns in the market, and delays or prevents the need to draw from the principal if the endowment earnings are negative.

Long-term management – ongoing

Tasks and costs are also created for every period following the initial period. Duties following the initial period are called 'ongoing' management. Thereafter, ongoing costs in the PAR model are performed on a repeated cycle – indefinitely. Obviously, some ongoing costs are incurred very frequently, such as biological monitoring and reporting, fence repair and other maintenance items. Others occur less frequently, such as replacing native plants lost to non-natural occurrences, and buying equipment or vehicles used in management duties. Other costs occur very infrequently, such as replacement of dikes and levees, water management systems or concrete structures. In essence, these are amortized costs.

Direct costs, indirect costs, overhead allocations and contingencies

After listing tasks, calculating hours and applying costs, I&C and ongoing costs are increased by a number of factors:

- indirect costs – employee benefits and payroll taxes;
- overhead costs related to office and administrative tasks including r&d, legal, executive, financial and personnel management;
- insurance, including hazard, liability, errors and omissions, fidelity bonds and officers', directors' and volunteer's coverage.

The sum of the direct, indirect, overhead and insurance costs is then multiplied by a factor, usually 10 per cent, which is set aside in the model as for contingencies – costs which may occur, but cannot now be quantified.

The PAR model adjusts all costs for inflation and discounts them for time. I&C costs are added together into one lump sum. In the best of all worlds that sum is collected at the time stewardship commences. It can be deferred and paid over the I&C period or as actually incurred. If I&C is deferred, it is advisable the steward receive some kind of guarantee or financial assurance of payment and an adjustment for Consumer Price Index (CPI) inflation and loss of endowment earnings. I&C

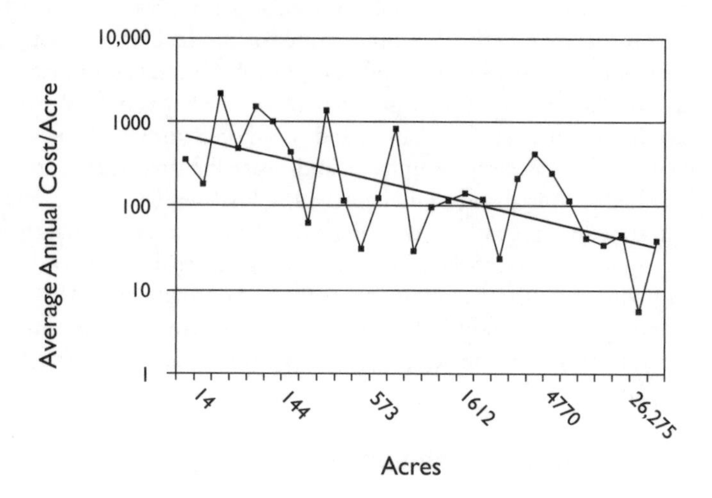

Figure 9.1 *Stewardship costs per acre*

costs may be collected from a number of sources, but it is most common, with conservation banks, to collect from the banker. Options are outlined below in the section on Endowments.

No average or typical costs – but costs decline as number of acres increases

The inflation-adjusted, discounted ongoing costs are also presented as a lump sum expressed in 'dollars per year'. PAR inputs include the number of acres in a project; therefore ongoing cost is also expressed as 'cost per acre per year'. This is helpful for some in making comparisons between projects. However, neither the resource agencies nor CNLM have ever been able to come up with a 'typical' or 'average' cost per acre for any habitat type or location.

Each conservation bank seems to be so site-specific as to defy categorization with any other. This appears to be a result of the proximity of most conservation banks to agricultural, ranching, timber, mining or urban uses. As a result, 'perimeter influences', such as invasive species, trespass, encroachment, exercise of easement rights by public utilities and similar site-specific impacts, dominate the ongoing tasks input into the PAR model. About the only generalization that can be made concerning stewardship costs is that ongoing management costs decline on a per acre basis as the size of conservation lands increases (see Figure 9.1).

Annual budgets for I&C and ongoing expenses

Obviously, ongoing annual cost is a combination of the cost of tasks performed repeatedly and those performed only sporadically over time. Because of this blending of frequently recurring costs and costs incurred only infrequently, a steward needs an annual budget. The budget line items mirror the I&C and ongoing task list, although they can be in greater detail.

The ongoing budget is calculated the same way as I&C – with a table of tasks, hours and costs. If a steward is relying on endowment earnings to fund ongoing expenses, three to five years of ongoing expense should be added to I&C costs and collected up front. This buffers the potential volatility in endowment earnings. The topic of endowments and their earnings will be discussed in the next section.

Budgeting and accounting for long-lived items such as equipment and improvements

The steward must have a means of accounting for long-lived items – equipment replacement, future construction obligations, etc. This would typically be done via a depreciation entry, where the depreciation amount coincides with the annual amounts estimated in the PAR model for needed equipment purchases or construction at the date in the future when they are required. The depreciation deduction is assumed to be reinvested at the PAR model's discount – or earning rate.

Managing shortfalls and adjusting tasks to meet changing conditions requires adaptive management. In the case of a decline in conservation values or failure to achieve stated objectives, management responses should assure eventual reversal of that decline or a return to a path that supports the original conservation goals. For a steward, these operating standards, together called 'adaptive management techniques', require flexibility in response to unanticipated events – changes in the land itself and the species that live there; changes in the neighbourhood and region surrounding the land; and changes in regulatory and land-use policy. In most cases, this kind of operational flexibility costs more than static management – in which a steward adheres to a fixed set of tasks. Even the most well grounded conservation plan should allow for change, contingencies and adaptation – and allow for shifts in tasks and increases in costs.

Sources of funding for management

Funding for annual I&C and ongoing costs can come from a variety of sources. Conservation banks have experimented over time with various mechanisms of assuring financing for critical tasks. These include upfront endowments, ongoing payments from conservation bank credit sales, grants, in-lieu fees and other sources of private or public financing.

Endowments

An endowment is broadly defined as money or property transferred to a holder (generally an institutional or non-profit entity of some kind) with the stipulation that it be invested and the principal, also known as the historic cost, remain unspent. Earnings from the endowment, which include interest, dividends and proceeds from the sale of appreciated assets over their historical cost (realized gains), are used to pay ongoing costs. Unrealized gains are also endowment earnings, but since the assets which have experienced gain have not been sold or converted to cash ('realized'), they are not available for spending – unless the steward borrows against the endowment in a margin account or otherwise. Such borrowing is generally prohibited in the documents creating or governing endowments.

A further stipulation by the parties establishing an endowment – or later stated in the written policies of those managing the endowment – is that the historic cost be augmented in such a way as to preserve forever its original purchasing power. That way earnings increase at roughly the same rate as ongoing costs increase – as a result of inflation. Unless there will be additions to the endowment by a banker or third parties, the only way for this augmentation to occur is to spend less for ongoing costs than the actual realized and unrealized earnings and gains on the endowment. Some portion of endowment earnings is not spent – it is reinvested in the endowment.

Currently, the US Fish & Wildlife Service encourages bankers to provide a monetary endowment to stewards sufficient to generate annual earnings equal to inflation-adjusted ongoing costs. They call this a 'non-wasting endowment'. That means the endowment's historic principal is not invaded and some portion of earnings is reinvested so that endowment principal grows, and earnings thereon are sufficient to fund inflation-adjusted ongoing costs in perpetuity.

Who invests the conservation endowment and why does it matter?

The right endowment for any given mitigation property is dependent on a number of complex evaluations. It is also conditioned upon whether the holder of the endowment is a public agency. With an endowment, a portion of the income from its investment is used for stewardship each year. The endowment can be a perpetual source of income for stewardship because, first, the principal is left whole, and second, the principal is allowed to grow at the rate of inflation. Only a portion of the endowment's income is used for stewardship each year because the remainder is added to the principal as an offset to inflation.

Most endowments for universities, hospitals and non-profits are invested in balanced portfolios. A portion is invested in bonds for security and short-term income and the remainder in stocks for appreciation. The division between the two is often 20 to 50 per cent bonds and 50 to 80 per cent stocks, depending upon the risk aversion of the organization and the stage of the business cycle. Depending upon the exact portfolio, over the long-term (say 30 years) returns from this kind of investment average between 9 and 9.5 per cent a year. Since inflation averaged about 4 per cent over the same period, there is, on average, about 4.5 to 5 per cent remaining for stewardship. However, governments lack this flexibility and are generally limited to lower returns. Section 53600 of the California Government Code requires that all investments by public or semi-public agencies in the State of California be limited to debt instruments.

A portfolio containing these investments generally returns 1.5 to 3 points less than more diversified portfolios. The lower return is caused by the restriction to debt instruments but also by the requirement that the issues have a very limited

Table 9.1 *Privately held endowment*

	Amount ($)	Per cent
Endowment	400,000	100.0
Investment earnings	34,000	8.5
Inflation reinvested	16,000	4.0
Stewardship income	18,000	4.5
Used for current expenditures and reserves		

Table 9.2 *Publicly held endowment*

	Amount ($)	Per cent	
Endowment	400,000	100.0	
Investment earnings	26,000	6.2	(Bonds only)
Inflation reinvested	16,000	4.0	
Stewardship income	10,000	2.2	
Used for current expenditures and reserves			

term. Although the reasons for the legislation are complex, it is generally designed to reduce losses of public monies by eliminating riskier investments. Unfortunately, Orange County, California sustained its dramatic monetary loss using repurchase agreements to bet on interest rates. Since governments generally are to receive public monies that are needed in the relatively short term for public expenditures, the law is also oriented to short-term funds. The endowment in perpetuity is a new, very foreign animal in public finance.

The point for conservation bank endowments, however, is that those held and invested by public agencies must be larger, about 100 per cent larger, than those held and invested by private entities in order to produce the same level of income for stewardship. It is a sensitive point but one that must be dealt with realistically.

Inflation and endowments

Inflation rates differ from task to task within an ongoing budget. Over time, employee benefits, health insurance, general insurance and vehicle costs have increased faster than the general rate of inflation as measured by the CPI. Periodically, energy costs increase faster than the general CPI, but from time to time they fall.

A steward should evaluate its task-based costs and, to the extent line item cost inflation has not tracked the general CPI, modify the PAR inflation assumptions based on the proportions of task-related cost in the various categories – those experiencing inflation at rates greater than the CPI, and those experiencing less cost increase, or actual decreases in cost, for example, for computer equipment.

Since the largest percentage of most ongoing costs are related to direct labour rates, and those rates have generally moved up at about the rate of inflation, as measured by the CPI, some stewards have not seen a pressing need to increase the inflation rate assumption in the PAR above the historic rate of change in the CPI. Of course all of this assumes a linearity and constancy of inflation and earnings rarely present in the real world.

There are risks involved with endowments, payments from conservation bank credit sales, bonds, grants, in-lieu fees and other sources of private or public financing. Extreme care must be exercised in any instance where the steward is not holding a sufficient endowment to ensure payment of inflation-adjusted costs on a timely basis. I&C costs are generally funded by direct payments. They are usually not funded from endowment earnings. As mentioned above, I&C should include three to five years of ongoing expense, to buffer the endowment from swings in earnings.

Determining the amount of an endowment

In all likelihood, a banker will want to keep the endowment as small as possible, while the long-term steward will want it to be as large as possible. The PAR provides a reasonable method of pinpointing the correct money amount of the endowment. The PAR inputs – tasks, hours and costs – can be listed and reviewed by the parties for necessity, accuracy and reasonableness. Bankers may also suggest deferring their contribution to endowment. They may believe it possible to augment endowment funding with grants, donations, fees and other sources of revenue.

Risk analysis, as suggested above, allows the parties to evaluate the amount, longevity and stability of sources of funding other than from endowment earnings. Bankers should note that the more complex a funding scheme, the more likely a steward will turn to legal and financial advisers for advice. These costs will of necessity be borne by the banker, along with commensurate delays in completing management and funding agreements and likely requirements of the steward for financial guarantees or assurances of payment. Endowment funding is, in most cases the most efficient, cost-effective and pragmatic way to assure long-term management of a conservation bank. Properly structured, endowment funding can relieve the banker of ongoing risks and long-term management responsibility for completed conservation banks.

Importance of income tax exemption

It is important to note that the owner and beneficiaries of an endowment should be a non-profit entity, not subject to local, state or federal income taxes on the receipt of the endowment from the banker, endowment earnings (interest, dividends and capital gains), endowment distributions, or grants, contributions and other payments from third parties in support of conservation management goals and tasks. Otherwise, the endowment and its earnings would have to be evaluated on an after-tax basis. That would greatly reduce the amount of money available to pay stewardship costs.

Bankers and stewards alike should be aware that not all forms of income received by non-profit entities are tax exempt. Revenue is subject to the 'unrelated business income' test of the tax code to determine if it is tax exempt. To the extent these entities also hold and manage endowments, endowment earnings can be exempt from tax. More recently, community foundations have been appointed as custodians or trustees of endowments, investing the endowment and paying earnings to stewards for ongoing expenses. It is beyond the scope of this chapter to deal with unrelated business income or income tax considerations of non-profits, but parties to a conservation bank should consult with counsel skilled at evaluating the tax implications of a land acquisition and long-term stewardship programme. This is strongly advised if the non-profit is also the bank sponsor and will realize income from the sale of credits.

Income capitalization

The PAR model provides one way to set the amount of endowment required to fund ongoing stewardship. The method is called 'income capitalization'. Capitalization is the process whereby periodic payments are converted into a single lump sum.

For example, if you need $10,000 per year for stewardship and can earn 4.5 per cent (after inflation) on an investment, the capitalized value of the income stream is:

Table 9.3 *Estimating the amount to invest*

Need to earn annually	$10,000
Capitalization rate	4.5% (Cap rate is a divisor)
Solve: $10,000/.045 = $222,222	
Amount to invest = $222,222	

Example 1

If budgeted spending is $10,000 per year and you can earn 4.5 per cent today on an investment in long-term US government bonds, how much money needs to be invested to fund your budgeted expenses?

On the face, that is the amount of the endowment. Every year, the 4.5 per cent earnings are withdrawn from the endowment and fund ongoing expenses.

$222,222 × 0.045 = $10,000
The $222,222 principal balance *appears* unchanged over time.

Impact of inflation on endowments and earnings

If ongoing costs rise over time, the second year's budget is going to be higher than the first. Over the past 70 years, the annual rate of inflation as measured by the CPI has averaged about 3.9 per cent. Here's an example of inflation over just the last 37 years:

1970 Price Index	
Gasoline	$0.36/gallon
Median Income	$8734/yr
Median Rent	$108/month
Median Home	$17,000
Bread	$0.24/loaf
Eggs	$0.51/doz
Harvard Tuition	$2600/yr
Harvard Tuition 2007?	>$46,000/yr

Earlier it was noted that the inflation rate for costs related to conservation lands may increase more rapidly. But in this example, we will use the long-term average rate of inflation to evaluate the adequacy of an endowment to pay ongoing expenses that increase over time due to inflation.

Year 2 budget will be $10,000 × 1.03 = $10,300
But the endowment will earn only $10,000 ($22,222 × 0.045)
Endowment earnings are $300 less than expenses.

What are the options?

1 Cut expenses.
2 Withdraw $300 from endowment principal and spend it.

In response to the first option – cut costs – the parties, using the PAR model, agreed on the tasks required to preserve conservation values. It is not rational to reduce or eliminate tasks everyone agreed were required to maintain property.

The necessity of tasks, their frequency and their cost is hashed out among the parties, including banker, steward, consulting biologists and regulators during the PAR process. But if $300 is withdrawn from endowment principal, endowment earnings in Year 3 will be ($222,222 – $300) = $221,922 × 0.045 = $9986

If inflation continues, Year 3 management cost will be $10,000 × 1.03 = $10,300 × 1.03 = $10,609. Only $9998 will be available from earnings. Further withdrawals from the endowment rapidly erode the endowment principal. In order to keep up with inflation in ongoing costs the endowment will be reduced to zero. In fact, this will occur in 28 years. The endowment will be completely used up. There will be no more funding for ongoing expenses. Neither option for managing expenses and endowment is feasible unless somebody will guarantee another large deposit 28 years down the line – reinvested at the rates then available – and on and on forever.

Transferring risk – impacts of inflation

In most conservation plans, the banker wants to acquire land, achieve success criteria, fund the endowment, sell credits and be done. The steward assumes long-term management responsibility and liability. Without an assurance of payments into the future, no steward would accept an endowment that could eventually be

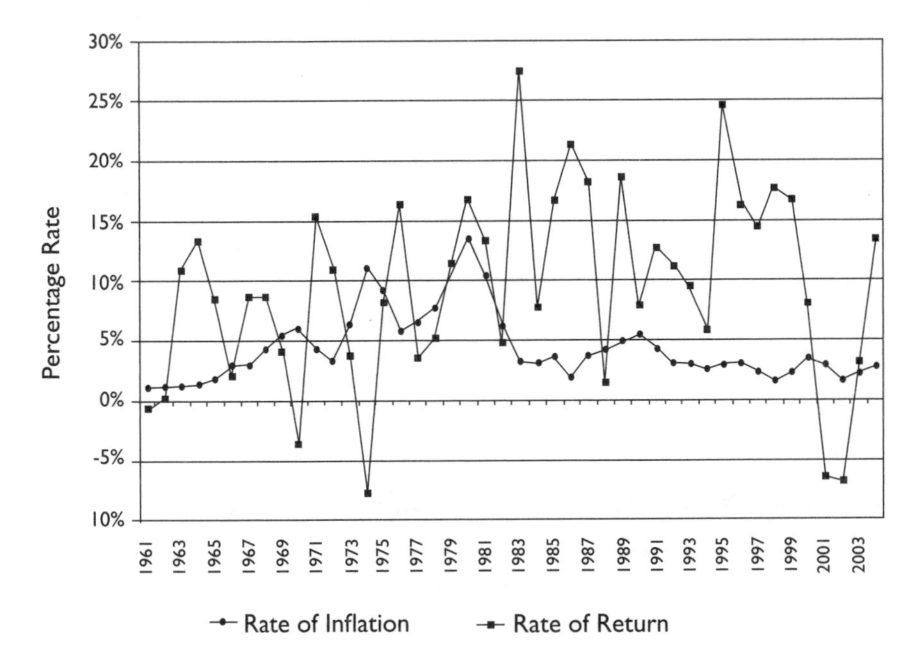

Figure 9.2 *Comparison of rates of inflation and investment return*

Note: Consisting of 60% stocks and 40% bonds.

zero. How would the steward pay costs associated with the land? As discussed, most conservation banks are impacted by development pressure and urbanization. There is no data from any source that suggests conservation banks can ever become permanently self-sustaining natural systems. They all seem to require stewardship. And stewardship costs money.

Looked at another way, it is possible to see how inflation erodes the buying power of an endowment. US government bonds pay interest every year. Principal is repaid when the bond matures. The longest term available today on a US government bond is 20 years. This is called its 'maturity'. Assuming 3 per cent annual stewardship expense inflation, the buying power of $222,222 invested in a 20-year bond would have a value in 2026 of just $120,086. The endowment will have lost 45 per cent of its *original* purchasing power. In 2026, assuming reinvestment of principal in another 20-year government bond at 4.5 per cent, on an inflation-adjusted basis, annual earnings in today's dollars would not be nearly enough to pay ongoing expenses of $10,000. In fact, the endowment would be completely out of money in its 28th year.

Nominal and real rates of return

Economists call the actual earnings rate of an investment, unadjusted for inflation, its 'nominal rate'. Deducting the effect of inflation from the nominal rate leaves us with what economists call the 'real' earnings or 'real' rate of return of an investment. Over time, the real rate of return on US government bonds ranges from about 1.5 per cent for short and intermediate term bonds to 4.5 per cent for long-term bonds.

Example 2

Therefore, if an endowment were invested in a pool of government bonds, and the endowment had to provide $10,000 real income every year, the endowment would range in size from $10,000/0.015 = $666,666 to $10,000/0.045 = $222,222.

This analysis assumes no investment management costs. A larger, more complex endowment holding multiple assets, and buying and selling to provide funds for ongoing costs, will incur management expenses. Those expenses would have to be deducted from endowment earnings or added to ongoing costs in calculating the endowment's initial principal balance. A $666,666 endowment is a far cry from a $222,222 endowment.

How to decide which endowment amount is correct?

A banker cares a great deal about which capitalization rate is used. The banker must come up with the money to fund the endowment. A banker would also like to reduce its ongoing risk in a project, especially once all credits have been sold. A steward cannot accept an endowment that is too small to generate enough earnings to pay ongoing expenses. The value of longer term bonds is highly sensitive to changes in interest rates. Shorter term bonds are less sensitive, but maturing principal must be constantly reinvested in order to fund ongoing costs. Therefore, even an endowment invested in US government bonds has risk. This is not the risk that principal or interest will be paid on time. This is the risk that earnings will fluctuate based on reinvestment rates and inflation will eat away at the purchasing power of investments over time.

Somewhere between the poles of low rates of real returns and too much risk lies the correct expectation of a sustainable real rate of return on endowment. Following the M&M period and when all credits have been sold, most bankers expect to transfer all future risks associated with a project to a steward. The steward should not accept a risk that its funding source is inadequate to pay all future costs as they are incurred. Therefore, even if there is no precise way to estimate the amount of an endowment, a reasonable amount can be established using known facts about how to estimate endowment returns.

Annual Budget	Cap. Rate	Endowment
$20,000	1.0%	$2,000,000
$20,000	2.5%	$800,000
$20,000	4.5%	$444,444

Endowments do not necessarily earn the 'average' return from year one

For the purposes of calculating an endowment, in the beginning the endowment earns the nominal available rates of return on assets in which it can invest. The endowment cannot necessarily find investments that earn the equivalent of the long-run 'average' return. The endowment earns what it can on the investments it can actually make. Today's short, intermediate and long-term US government bonds all have nominal yields (earning rates) less than 5 per cent. Inflation is currently around 3 per cent. Today's investment in government bonds will return 2 per cent in real terms (5 per cent yield minus 3 per cent inflation). This is less than the long-term average return, but it is the best the endowment can earn unless it diversifies its investments. Barclays Global Capital Advisors have calculated percentage rates of return for US stocks and bonds from 1926 to 2005:

Stocks	9.8
Bonds	7.9
Money Market Funds	4.0
Adjusted for 3 per cent inflation, the real returns are:	
Stocks	6.8
Bonds	4.9
Money Market Funds	1.0

Endowments are created assuming investments in both stocks and bonds

To simplify the PAR model, and calculate the size of an adequate endowment, ongoing costs are capitalized by the average real returns available from a portfolio of stocks and bonds. The returns reported by Barclays are averages over a 79-year time frame. They are reported before investment management expenses. And, they do not take into consideration the likelihood that actual annual earnings will be more or less (the variance or standard deviation of returns). Assuming the endowment were invested 60 per cent in stocks, 30 per cent in bonds and 10 per cent in money market funds, it would be expected, over the very long run, to earn real percentage returns of:

Stocks	6.8 × 0.6 = 4.08
Bonds	4.9 × 0.3 = 1.47
Money Market Funds	1.0 × 0.1 = 0.10
Sub-total	5.65
Minus investment management costs	1.00
Expected annual real return	4.65

This long-term expected real rate of return is as reasonable a benchmark as any in estimating the amount of endowment required to fund ongoing costs.

Example 3

$10,000 ongoing cost divided by 4.65 per cent annual real rate of return equals $10,000/0.0465 = $222,222 endowment

An endowment invested in a diversified portfolio of stocks and bonds should return 4.65 per cent per year, after inflation and investment management expenses. But isn't that very close to the 4.5 per cent rate of return on bonds we assumed in Example 1? What has changed?

Example 1 capitalized ongoing costs at the nominal rate of return – before adjustment for inflation, and not including returns from assets – on bonds that repay a fixed principal sum at maturity. That fixed sum must be reinvested to provide earnings for future ongoing expenses. It is subject to losing purchasing power as a result of inflation.

In Example 2, when we adjusted Example 1 for inflation – and where the endowment was invested only in US government bonds – the required endowment ranged from $222,222 to $666,666.

In Example 3 the endowment investment portfolio consisted of stocks and bonds. The nominal return on this new portfolio is about 8.65 per cent. The real return, minus investment management costs, was 4.65 per cent.

Stocks earn more

Bond principal is subject to erosion by inflation, whereas stock values tend to rise (and fall) more in line with inflation – as both have similar roots. Importantly, investment in stocks provides more assurance that there will be returns, in excess of ongoing expenses and investment management fees, that can be reinvested in the endowment, compounding the principal value of the endowment, and assuring that tomorrow's inflated ongoing expenses can be paid out of returns – and not out of endowment principal.

Spending and reinvestment of endowment earnings

Therefore, in managing the endowment, two requirements must be observed:

1. Spending may not exceed a fixed percentage of the actual endowment principal earnings.
2. All earnings in excess of that amount must be reinvested in the endowment.

Note that spending from the endowment is based on expected endowment earnings, adjusted for inflation. In some years, actual returns will be less than expectations. In some years they will be more. Actual spending, if held *below* 5 per cent of the endowment balance in an endowment invested in a diverse pool of debt and equity instruments, is not likely to erode the purchasing power of the endowment.

Therefore, even if endowment returns are not 4.65 per cent in a given year, spending can be 4.65 per cent of the endowment balance. (See below for discussion on prohibitions against spending endowment historic principal, and policies related to investment management benchmarks and performance evaluation. See also above for recommended increases in I&C to account for variation of returns in the initial years of endowment investing.)

But, in years in which endowment earnings exceed 4.65 per cent, the excess must be reinvested. The documents transferring the endowment to the steward should allow the steward or endowment manager this flexibility – to spend a fixed percentage of endowment assets annually, and to reinvest all excess earnings.

Policies related to endowments

Any entity charged with management of endowment funds or their proceeds should have accounting and investment management skills and adopt written investment and spending policies. At a minimum these policies should include written descriptions of:

- endowment objectives;
- role of the steward, its staff, board of directors and consultants;
- selection of investment managers;
- risk management – using benchmarks and performance evaluations;
- payment of investment management fees and expenses;
- asset allocation;
- spending and reinvestment.

According to the Ford Foundation and others who have researched endowment investment and spending, two of these policies are absolutely critical to long-term endowment performance (Siegel, 2001):

1 Which investments to buy with endowment funds.
2 The rate at which endowment earnings are transferred to the operating budget.

These two decisions have more impact on the longevity of an endowment and its ability to pay ongoing costs than any others. The first decision, what investments to own, is called 'asset allocation policy'. Allocation policy is the optimum mix of investments to achieve a targeted rate of return while limiting risk. The second decision, what to spend, is called 'spending policy'. A complete analysis of asset allocation strategies and the financial theories and statistical methodologies used to formulate them needs to be carefully researched and evaluated. Unfortunately, there is no precise answer to the investment allocation. Commonfund reviewed investment returns for 2005 (Commonfund, 2006) and described asset allocations by the size of the endowment.

Spending policies

Data are available to allow stewards and endowment managers to make assumptions about the ability of endowment principal to retain its purchasing power over time based on asset allocation and spending decisions. Thus, if annual spending is set at 5 per cent of endowment principal, and equities comprise 80 per cent of an endowment's investment, according to the Ford Foundation there is a 68 per cent or better probability that an endowment's balance will keep up with inflation at least five years from the present. If spending is reduced below 5 per cent, there is a much greater probability that the endowment principal's purchasing power will stay intact over a longer time. Based on its 18 years of experience managing endowments, CNLM recommends establishing an endowment based on the capitalization of ongoing expenses at 4.5 per cent or less depending on the size of the endowment, and annual spending of endowment principal at an amount not in excess of 4.5 per cent.

Special resources

Commonfund and The Investment Fund for Foundations (TIFF) specialize in policymaking, implementation and performance measurement for foundations and others holding endowments. Both are themselves non-profit entities. Information on their services is available on the web at www.commonfund.org and www.tiff.org.

Endowments invested by public agencies

Many jurisdictions have no provision for sequestering funds dedicated to a specific parcel or parcels of land. At best, money is deposited into segregated accounts. But in many cases spending from those accounts must be approved, often annually by a legislative body. In other cases, money is deposited into the general fund and all expenditures must be appropriated, annually by a legislative body. Many jurisdictions have no lawful ability to invest funds in assets that have historically produced enough earnings to fund monitoring and management in perpetuity.

In California, the state treasurer manages a fund for deposits from various branches of government. Although inflation in the Los Angeles, California region has averaged 4.85 per cent over the past 35 years (CPI, All Urban Consumers LA Riverside Orange All Items) the average annual return on this fund has been 6.83 per cent (California State Treasurer and Pooled Money Investment Account/Local Agency Investment Fund Performance Reports, 2006) and in the past 15 years, the fund has earned just 1.72 per cent a year on average, versus annual inflation of 3.4 per cent. As a result, funds invested by the government to fund conservation would have experienced a real (inflation adjusted) return of less than 2 per cent per year over the past 35 years (6.83 per cent fund earnings minus 4.85 per cent inflation rate) – and a negative –1.3 per cent in the past 15 years. As we shall see in subsequent sections, without a very large endowment from which to draw funds for stewardship, this real rate of return is wholly inadequate to pay ongoing management costs and keep up with inflation.

In fact, to achieve the same goals (i.e. pay inflation-adjusted expenses), an endowment held by a public agency would have to be more than twice the size of one held by a private non-profit and invested in a broadly diversified pool of bonds and stocks. Unfortunately, the nation's largest municipal bankruptcy, in 1994 by Orange County, California, occurred as a result of its treasurer's attempt to earn higher yields on government funds entrusted to him via speculation in securities linked entirely to otherwise safe US government or agency bonds. There is no guarantee that public entities can own land and provide funding for conservation programmes with any greater assurance of success than programmes created by the private sector, or in public–private partnership.

While the resource agencies want the security of the endowment principal, they are unable to achieve it through investments in state funds. The federal government will not hold long-term endowments. In California, the situation is further exacerbated by the fact the state has not been able to make timely payments or any payments at all (one mitigation banking company has yet to be paid for stewardship from state-held endowments four years running) and California law now requires that payments from these endowment accounts go through the annual state budget process, which would even further delay payments to the bank steward.

This is why public–private partnerships are critical to establishing and maintaining successful conservation banks. The private conservation entity or a foundation may hold an endowment which can be prudently invested in a balanced, diversified portfolio of assets that earn substantially higher risk-adjusted returns than available to public agency treasurers. Finally, in the event the private entity is unable to fulfil or pay for its mission, the public can step in and 'cure' the problem with monitoring and management expertise and additional funds.

What happens when the bank sponsor fails to pay?

Such is the case with a combined mitigation and conservation bank in northern California. This example also highlights the need for guaranteed funding upfront and the hesitancy of bank stewards to accept phased payments for long-term funding. A signed agreement between the bank sponsor and steward stipulated that the endowment would be funded from credit sales but not exceeding five years. The I&C costs (annual management funding) would be paid in annual payments with the first year upfront until the endowment was fully funded. The initial payment was made and several endowment payments, but the annual stewardship funding is currently three years in arrears. This means that in order to maintain the site, withdrawals from the endowment were required. It is rapidly approaching Year 5, the endowment is unfunded, annual stewardship is reduced due to reduced funding, the endowment principal has been reduced and unaccounted hours of staff and legal time and resource agency time have been spent trying to get the bank sponsor to fulfil their obligations. If the endowment had been funded from the beginning, there was an unrealized gain during that period of over 40 per cent or over $450,000 that would have been added to the endowment from investment income. The bank sponsor currently owes over $900,000. Resource agencies are attempting to work with the banker to get payments current, or they will suspend credit sales. However, the agreements and contracts do not stipulate that the banker pay for investment income lost during the late payment period.

Establishing accounts

Generally, each individual conservation bank will have an associated endowment and ongoing spending programme. The steward and/or endowment manager must have a way of tracking each project's annual budget, original PAR cost estimates and the timing of capital expenditures, and associated endowment earnings and principal balances.

Fund accounting and concentration accounts

Each project should have its own budget, balance sheet and income statement. Periodic analysis should be performed to evaluate variances between budgeted and actual expenses and anticipated and actual endowment earnings and inflation-adjusted endowment principal balances. On the other hand, it is highly impractical for stewards and managers of multiple projects to establish separate banking and endowment investment accounts for every project. Most modern finance and accounting programmes allow for this kind of 'concentration' of banking and investments, yet retain budget, balance sheet and income statement detail at the project or property level. This is a process called 'fund accounting'. Fund accounting principles should be used in conservation banking stewardship.

Prudent investor rule, and the Uniform Management of Institutional Funds Act

To the extent an endowment is held in a trust, it is governed by state laws, including the 'prudent investor rule'. The trustees of a trust have a duty to the beneficiaries to invest and manage the funds of the trust as a prudent investor would, in light of the purposes, terms, distribution requirements and other circumstances of the trust. The law changed in 1992 to broaden the concept of a 'prudent man', whose actions with regard to investing for others was defined as '... how men of prudence, discretion and intelligence manage their own affairs, not in regard to speculation, but in regard to the permanent disposition of their funds, considering the probable income, as well as the probable safety of the capital to be invested.'

Even if an endowment is not held in trust, there is a uniform state statute that deals with how it should be managed and invested. This is called the Uniform Management of Institutional Funds Act (UMIFA). It has been adopted in similar form by the states and a revision of UMIFA is in progress.

Accounting, audits, reporting and taxes

Stewards must have a means of keeping track of assets and liabilities, estimating costs, creating budgets, paying bills, evaluating spending, measuring endowments balances and returns, and forecasting whether their financial resources are adequate to protect the conservation values. This means there must be an accounting system in place. Stewards and investment managers should use generally accepted accounting principles (GAAP), consistently applied. This enables a third party auditor to review financial performance and operating policies and procedures with respect to finance and report on the accuracy of financial statements and the sufficiency of financial risk management.

Some agreements will require bankers, stewards and investment managers to prepare annual audited financial statements. Internal Revenue Service filing requirements with respect to transfer of land, bargain purchase of land and other contributions of real and personal property and cash related to conservation banks can be exceedingly complex and a banker should seek the appropriate professional advice. Regarding bankruptcy of conservation banks the reader is directed to Gardner and Radwan (2006) and Teresa (2006).

Credit registry and master escrow account

Wildlife agencies are encouraged to maintain a regional credit registry (such as USACE's Regional Internet Bank Information Tracking System [RIBITS]). The registry will note the initial number of credits and their purpose; any additions, such as creation of new credits for multi-species banks, additional enhancements, and annexation of new phases; and all sales or uses of credits.

All bank sales transactions would be performed by a nationally recognized escrow company. The master escrow holder may be a title company escrow or other neutral, independent, third party with expertise in permits and documents related to land and financial transactions. While the actual price of the credit could remain confidential, all parties would be alerted to the sale, with the number and type of credits being accounted for and all relevant parties would receive compensation. This is especially true of the habitat steward who may be funded over a period of time through the sale of credits.

In addition to the conditions of purchase and sale contained in the purchase agreement, the master escrow will require the following:

- approval of the sale/use of credits by appropriate permitting authorities;
- entry of the sale or use in the credit registry as 'pending';
- search for liens against conservation lands or conservation credits and collection of funds or alternate security to release liens;
- beneficiary statement from the bank's steward stating the amount of i&c or endowment, if any, due from the credit sale;
- method of collecting and disbursing funds.

The purpose of the registry and escrow is to systematize credit tracking, prevent 'double-dipping' (i.e. double selling) of credits and assure all parties that the conservation values supporting the credits are intact.

Conclusion

A conservation bank's biological resources and stewardship funding must be protected in perpetuity to fully mitigate for the loss of special status species. The conservation values of a bank can be permanently protected through the use of conservation easements, even for lands in public ownership. The PAR provides stewards a defensible way of determining what stewardship tasks and their attenuating costs are for the long-term protection of the conservation values. Sources of funding for acquisition, restoration and long-term management must be available and assured – in some reliable form – 'upfront'. Payments to an endowment or ongoing stewardship can be deferred through credit sales, but security must be given for that deferment – preferably in the form of financial guarantees, but in all cases with the certainty of enforcement. The permitting agency should undertake enforcement when agreements are violated.

Funding must keep up with inflation, either by continuous contribution of funds from a reliable source, earnings from endowment, or a combination. The amounts available to pay for long-term management tasks must not be fixed. They must move up at least at the rate of inflation of the costs associated with those tasks. The best way to assure this is the investment of a sufficient sum, determined by capitalizing estimated annual ongoing management costs at a rate not in excess of 4.5 per cent (based on historical real earning power of investments since 1926), in a diversified portfolio of stocks and bonds. Endowments must bear risks in order to earn adequate returns. Funds must be managed, invested and accounted for using standard rules, recommendations and reporting formats.

There are four financial aspects to ensure a sound financial footing for conservation banking. These are:

1 Fully document the tasks associated with long-term management and quantify their current costs.
2 Set aside an endowment sufficient to pay those costs, accounting for the effects of inflation. Allow NGOs to invest these funds under strict standards and guidelines and with consistent oversight from the regulatory agencies.
3 Integrate documents, track credit sales through an escrow and provide public agencies the wherewithal to enforce them quickly.
4 Monitor and track the overall success of compensating for loss of species and habitats through the use of conservation banks.

Regulators and public agencies must have confidence in bankers and stewards, and may reasonably request some kind of guarantee of performance, so written agreements must bind all parties and all parties should have a place at the negotiating table and documents should be integrated and cross defaulted. With all parties working cooperatively, conservation banks can be a significant means of achieving species conservation and, ultimately, recovery.

Acknowledgement

With much gratitude to Peter St Clair for his valuable assistance and guidance.

Bibliography

California State Treasurer's Office (2006) 'Pooled money investment board fiftieth annual report fiscal year 2005–2006', www.treasurer.ca.gov/pmia-laif/reports/50annualrpt.pdf, accessed 6 July 2007

Commonfund (2006) 'Private equity, hedge funds, and other alternative assets grow to almost one-quarter of holdings of 60 percent of institutions surveyed', www.commonfund.org/Commonfund/Archive/CF+Institute/CBS_educational_press_0106.htm, accessed 17 April 2007

Compass Point Nonprofit Services (2006) 'What is the board's responsibility in investment?', www.supportcenter.org/askgenie/details.php?id=82, accessed 6 July 2007

Environmental Law Institute (2002) 'Banks and fees: The status of off-site wetland mitigation in the United States', www2.eli.org/pdf/d12_08ExecSumm.pdf, accessed 26 March 2007

Fox, J. and Nino-Murcia, A. (2005) 'Status of species conservation banking in the United States', *Conservation Biology*, vol 19, no 4, pp996–1007

Gardner, R. C. and Radwan, T. J. P. (2006) 'What happens when a wetland mitigation bank goes bankrupt?' *National Wetlands Newsletter*, July–August 2006, vol 28, no 4, pp16–21

Ibbotson Associates (2006) 'Ibbotson asset allocation methodology', http://corporate.morningstar.com/ib/documents/MarketingOneSheets/AssetAllocation_102506.pdf, accessed 10 July 2007

Siegel, L. B. (2001) 'Investment management for endowed institutions', www.fordfound.org/publications/recent_articles/investman.cfm, accessed 24 June 2007

Teresa, S. (2006) 'The demise of the environmental trust', http://ecosystemmarketplace.com/pages/article.opinion.php?component_id=4227&component_version_id=6060&language_id=12, accessed 1 July 2007

US Department of Commerce, Bureau of Economic Analysis, 'Gross state product data table', *Regional Economic Accounts*, 26 October 2006

Part III

State of the Art

10

Fish Banking

Tom Cannon and Howard Brown

Introduction

Fish mitigation and conservation banks are a relatively new and upcoming type of banking. Two fish banks have already been developed under the US Army Corps of Engineers (USACE) mitigation banking programme and the federal conservation banking programme, and several more are in the planning stages. As for other bank types, fish banks satisfy mitigation requirements of the Clean Water Act (CWA) section 404 and can be applied to meet the purpose and requirement of the Endangered Species Act (ESA).

The first fish bank was the Kimball Island mitigation bank, a USACE-entitled mitigation bank in California that included 'fish credits'. The fish credits were 'Chinook salmon' and 'delta smelt' habitat credits authorized by the National Oceanic and Atmospheric Administration's National Marine Fisheries Service (NMFS) and the US Fish and Wildlife Service (USFWS) (otherwise referred to as the Services) and approved by the USACE Sacramento district mitigation bank review team (MBRT). The Kimball Island bank is a 109-acre restored tidal marsh located in the Sacramento–San Joaquin Delta in south-western Sacramento County north of the City of Antioch, California. The bank was approved in 1998 and 'sold-out' in 2006. The majority of credits sold were prescribed as mitigation for development projects affecting the federally-listed delta smelt.

The second fish bank, the Fremont Landing salmon conservation bank, was entitled in 2006 by NMFS under the federal conservation banking guidelines. The Fremont Landing bank is the first fish conservation bank entitled by NMFS under its own banking programme. The bank is a 100-acre river floodplain site on the Sacramento River approximately 20 miles north of the city of Sacramento. It has credits for sale for three federally-listed salmon populations and their critical habitats: Sacramento River winter-run chinook salmon, Central Valley spring-run chinook salmon, and Central Valley steelhead[1] and the threatened southern distinct population segment (DPS) of North American green sturgeon. Credits also are available to compensate for effects on non-listed anadromous fish and their habitat, including the essential fish habitat (EFH) of Pacific salmon designated under the Magnuson-Stevens Fishery Conservation Act (MSA), as amended.

In this chapter we provide background on fish banks, how they are being developed, problems being encountered, and what potential roles they may play in recovery of listed fish species in the Pacific north-west and California.

What are fish banks?

Fish banks are mitigation or conservation banks that involve listed fish species and their habitats. Fish banks are no different than other types of wetlands or species banks used for mitigation under sections of the CWA or ESA. Under USACE guidelines mitigation banks should focus on watershed function such as special-status species habitat. Under the ESA, conservation banks are used

> *...when on-site conservation measures are not practicable for a project or when the use of a bank is environmentally preferable to on-site measures... From the Service's perspective, conservation banking can reduce the piecemeal approach to conservation efforts that can result from individual projects by establishing larger reserves and enhancing habitat connectivity. (USFWS, 2003)*

This guidance is especially appropriate for fish banks because fish habitat mitigation can be difficult to accommodate within a development site, and because fish species generally require large connected habitats to fulfil their life history requirements.

Benefits of fish banks

Fish banks are an effective means, if aptly applied, to provide mitigation that aids in the recovery of listed fish. Banks can provide strategically placed and sized habitat within a listed species range. The approach fits well with the Services' approach of characterizing and listing a species at the scale of ecologically significant units (ESU) or distinct population segments (DPS). An ESU or DPS is a population of organisms that is considered to be distinct for conservation purposes. Often times, geographic isolation is a major consideration in designating an ESU or DPS. Well planned banks can also provide habitat when needed so there are no lags or shortages of habitat between impacts and mitigation. Perhaps most important to recovery, banks can provide an immediate benefit by protecting existing quality habitats or adding substantially to quality habitats available. Banks also offer assurances that the properties and habitats will be protected, restored and managed in the future.

Because fish habitat banks offer these advantages, credits from fish banks can be very effective conservation measures when offered by a project applicant. With bank credits (habitats) pre-approved and guaranteed, the severance of liability to both the applicant and the permitting agency can be reassuring.

Banks also help all parties involved in development actions to meet their responsibilities under the ESA (section 7(a)(1)) to undertake actions that help conserve threatened and endangered species. USACE and the Services use banks for this purpose in their permitting programmes. The ESA specifically mentions activities such as 'habitat acquisition and maintenance' with lands acquired by 'purchase, donation, or otherwise' consistent with conserving the species or their critical habitat.

Other benefits of fish banks include: (1) providing a testing ground for new techniques through adaptive management; (2) bringing private investment to recovery programmes;[2] and (3) increasing the amount of economic benefit of recovery efforts reaching local landowners and communities. Testing and adaptive management are necessary given the technical uncertainties in restoration programmes. Private investment is important because it brings diversity to an area that is often dominated by public funds, with mixed or uncertain results. Also, funds from regional grant programmes have a tendency of not reaching local levels, whereas fish banks can be developed in partnership with landowners or communities looking for ways to supplement their income while protecting and restoring habitats on their conservation lands. More of each restoration dollar also stays in the local economies. Fish banks are a viable option for the 'community-driven recovery coming from self-generated economic activity' that Bailey and Boshard (2006) describe as an important potential tool in the recovery of Pacific salmon.

Role in species recovery

Can fish banks actually contribute to recovery or are they simply small potatoes in the overall mitigation and recovery programmes? The answer to this question depends on two primary factors. First, whether there is sufficient demand for off-site mitigation in service areas to support fish banking programmes. Second, whether fish recovery programmes rely primarily on large-scale habitat conservation plans (HCPs) or grant-funded restoration programmes (e.g. California's CALFED program[3]). Salmon recovery programmes are often well funded with hundreds of millions of dollars of taxpayer commitment. At a minimum, the role of fish conservation banks will be to provide effective mitigation for a myriad of small development projects that are not associated with the larger water supply, flood control, transportation, port and other public works projects. At a maximum, fish habitat banks will be a primary source of mitigation for large development projects and key elements in recovery programmes. Whether there will be sufficient stimulus for private funding of fish habitat banks will depend primarily on market demand.

With demand being important, the location of the early fish banks has typically been driven by the market for credits. Banks were developed close to project impacts. Although this is a logical approach based on the concept of in-kind and on-site mitigation, given the highly migratory nature of many of the listed fish species, this approach may not provide the greatest benefit to the listed species. Therefore, an important consideration for fish banks will be locating them in areas that are biologically significant for the species. This requires fish bankers to consider concepts such as species viability and the regional recovery needs. To accomplish this, banks must contribute towards alleviating the ongoing or proposed threats to species viability regardless of the specific demands for mitigation.

In 2000, NMFS developed the viable salmonid population (VSP) concept as a basic framework for assessing population viability (McElhany et al, 2000), and has since applied this concept in recovery planning. Under the VSP approach, anadromous fish populations are screened against four basic viability parameters: (1) population abundance; (2) population growth rate; (3) population spatial structure; and (4) diversity (both genetic and geographic). Conservation banks should be developed with these parameters in mind. The following are some examples of how to integrate these concepts in conservation banking.

Population abundance

This parameter considers population size. In general a species with a low population faces greater risks than a large population. Banks can address this parameter through the sale of credits that enhance and create new habitat, and by strategically locating bank sites in biologically meaningful areas that leads to increases in population abundance. Habitat enhancement and creation credits influence population abundance by improving local growth and survival conditions, which, in turn, can increase the production capacity of an area. Individual bank sites can be strategically located to benefit the greatest number of individuals, or the greatest number of populations. For example, a bank site placed at the confluence of two rivers would be expected to affect more individuals and populations than if it were located upstream on one of the tributaries.

Population growth rate

Similarly, habitat protection and enhancement at fish conservation banks can improve local growth and survival conditions and thus improve the overall fitness of affected populations. In order to have this effect, fish banks must be of sufficient size that the fish using the banks will benefit from their exposure to good habitat conditions, and banks must be located in areas that will be used by a significant number of individuals in the population.

Population spatial structure

This parameter considers the spatial structure or geographic distribution of a population. Spatial structure is important because it affects evolutionary processes and may alter a species' ability to respond to environmental change. Conservation banking considers this parameter by providing a broad geographic structure that considers the entire range and distribution of a species. In conservation banking programmes individual bank sites can be located and tailored to fit the needs of a species and its life-cycle requirements throughout its range.

Diversity

Many species exhibit diversity within and among populations. Genetics is a critical element of this parameter, but diversity also includes others factors such as variations in migration timing, habitat use, behaviour and size. Many of these diversity traits reflect variable environmental conditions that affect the species over time. Conservation banking can incorporate these elements by developing habitat features that are available to fish under a variety of environmental conditions. For example, a salmon conservation bank may include preservation, enhancement and creation of floodplain habitats that are inundated by high river flows to provide rearing and refugia for emigrating juvenile salmon in wetter seasons. The same bank also may include nearshore habitats that protect or improve rearing and refugia conditions for juvenile salmon of other races that emigrate during other seasons that have lower flow and less floodplain available.

Why have fish banks been slow to catch on?

One reason why there have been few mitigation and conservation banks to offer fish credits is the generally held position that the use of banks should be applied only to projects with small impacts and that larger impact projects should be self-mitigating. A second reason is that public agencies in need of mitigation generally believe it is less costly to self-mitigate than to invest in conservation banks. Thus with limited demand there has been little need for fish banks, and no new fish banks have been entitled by the USACE mitigation banking programme after the Kimball Island mitigation bank. The Kimball bank did eventually sell out in 2006. The slow sales history and a continuing trend towards project-specific mitigation has contributed to private investment dollars remaining focused on project-specific mitigation and development of non-fish species conservation banks.[4] But past trends are beginning to change and fish banking is expected to increase in the coming years, as discussed below.

New interest in fish conservation banks

The notion that conservation banks can play a significant role in the recovery of listed fish species is creating new interest in fish conservation banking. Two regional NMFS offices, south-west and north-west, are now involved in conservation bank entitlement, and see conservation banks as potential conservation measures in permitting small as well as large development projects. Large projects include major federal–state flood control or water supply infrastructure projects that often have multiple small elements over a wide regional area with phased implementation. Because impacts of such projects are often cumulative over time to fish populations, fish banks make an ideal form of conservation measure. The Services have begun working with the Oregon Department of Transportation (ODOT) and the Oregon Department of Fish and Wildlife (ODFW) to develop fish banks. They have developed management plans for habitat protection, habitat improvement and monitoring fish population and habitat trends at ODOT properties. The first two fish banks will be certified in 2007.

The success of the USFWS conservation banking programme has also generated interest in fish banks. For some years now USFWS has recognized that public and private developers are hard pressed to plan, design, construct and manage their own mitigation, and thus strongly supported use of conservation banks to provide maximum benefits to the species while not hindering timely well-planned development. Conservation banking also offers the Services more direct control of the mitigation process as well as providing a direct link between mitigation and recovery programmes.

The traditional approach to mitigation of large-scale public works programme developments has been project-specific mitigation and grant investments in conservation and restoration, which have had limited success in helping towards the recovery of listed fish. Poor performance for some of these mitigation projects has not helped. In contrast, conservation banks have performed well under ever-increasing strict entitlement and management rules that provide assurances that habitats will be built, managed and protected in perpetuity.

Umbrella fish conservation banks

Umbrella bank agreements (UBAs) are a potentially valuable tool in the implementation of fish conservation banking programmes. A UBA is simply a mutually agreed upon set of procedures and policies for entitling conservation banks between the Services and the bank developer. A UBA includes a set of mutually agreed upon rules and stipulations under which a bank developer and the entitling Service locate, entitle, design, build and manage mitigation or conservation banks. With NMFS, umbrella agreements focus on developing conservation banks for specific ESUs and towards meeting VSP goals for the ESU. Because NMFS addresses listed fish by ESU, often with specific recovery plans and VSP goals for distinct populations or population groups, umbrella banks become part of the recovery programme. UBAs also help ensure fish bank developers follow the lead of the recovery programme.

The focus of fish recovery programmes being generally on populations or ESUs within specific watersheds lends itself to the umbrella bank process. Both Services' listing approaches also lend themselves to the umbrella banking approach wherein conservation banks are planned with one overall agreement for each listed species or population, or recovery plan (if available).

UBAs include the terms and conditions for establishing individual conservation banks. These become the guidelines for banking in a specific ESU and include requirements for individual bank entitlements called conservation bank agreements (CBA). A UBA defines how individual banks are to be established, what the service area of the banks will be (usually the ESU or a watershed boundary), the form and content of individual bank agreements (including habitat development and management plans) and any limitations. The UBA also provides definitions, outlines procedures and rules for establishing easements, endowments, financial accounts, any related trust agreements, credit release criteria and schedules, credit accounting procedures, general monitoring and management procedures, agreement termination criteria and bank closure rules. The UBA may even address the process of obtaining federal or state permits including procedures for section 7 consultations for bank permitting. UBAs may include a management plan that details how bank sites and properties will be managed to protect and improve habitat, and to measure and monitor performance. UBAs can also prescribe content for bank-site specific habitat development and management plans.

UBAs can also deal with the ultimate disposition of conservation banks and how they will be managed and participate in the long-term recovery programme for the species. Rahr and Augerot (2006) recommend the 'creation of an area-wide network of salmon sanctuaries for selected stocks and populations and transferring ownership of the land and associated water licenses to non-profit salmon societies' (Lackey et al, 2006). Conservation banks and UBAs could be excellent tools for establishing such sanctuaries.

The use of umbrella fish conservation banks provides for the improved focus on species recovery. The NMFS Sacramento office has initiated an umbrella Central Valley salmon conservation bank programme that ties future conservation banks in this watershed directly to the species recovery process. Fish habitat bank sites will be developed only after consideration of the species recovery needs as well as future mitigation needs. This way money and effort are allocated where they do the most good for the species. The umbrella conservation banking programme also includes monitoring, research and adaptive management in the entitlement of individual banks to maximize the future effectiveness of the programme and to further contribute to the conservation and recovery of the listed species.

The Central Valley UBA also includes a conservation plan that details how conservation banks can contribute to meeting VSP goals and the recovery of the target species or population. The conservation plan details how sites are chosen, habitat type and location priorities, and what areas are in need of restoration. The conservation plan makes specific reference to applicable recovery plans, VSP goals and other related plans addressing the species or populations targeted by the UBA. The conservation plan outlines preservation, restoration and creation approaches/alternatives that banks may employ for specific circumstances and identify measures the bank developers will employ to ensure the umbrella bank's commitment to the targeted species or population recovery.

The broader focus of umbrella conservation banks also allows conservation bank programmes to better address the mitigation needs of large-scale public works projects and ensure that mitigation dollars are used to the maximum benefit of listed species. One common characteristic of public works programmes is that they tend to stall or become delayed from lack of funding or other reasons. A mitigation programme implemented with an umbrella conservation banking programme can be effective in addressing such uncertainties. For example, private funding can help to fill in periods when public funding is unable to sustain planned mitigation.

For the large public works programme, umbrella banks can be a cost-effective tool in addressing future mitigation needs. The ODOT fish banks mentioned earlier are a good example of a public-supported umbrella banking programme. The two banks being entitled are under ODOT's Statewide Mitigation/Conservation Agreement, an umbrella-type agreement, with state and federal agencies. The agreement includes bank management plans and credit and accounting systems.

Remaining stumbling blocks for fish habitat banks

Under the CWA, USACE policy is 'no net-loss of wetlands' that includes a reluctance to allow species preservation or restoration credits as compensation for the loss of wetlands habitats. Unlike wetland mitigation banks, species mitigation under ESA may include preservation and restoration of existing habitats. The dilemma comes when project impacts are to aquatic habitats that come under the jurisdiction of the CWA and ESA. The Kimball Island mitigation bank, entitled by USACE and the Services as a combination CWA and ESA bank, provided both fish credits and wetland credits. In contrast, the NMFS-entitled Fremont Landing conservation bank has no wetland credits as USACE did not participate in the entitlement process. Entitling fish species banks in the USACE wetland mitigation banking programme has proven difficult and very time-consuming, primarily because there are many federal, state and local agencies invited to participate in the process and it is difficult to address all their issues and wishes, and obtain entitlement in a timely manner. The process has proven so cumbersome that it has caused the Services to develop fish banks under conservation banking programmes outside the USACE mitigation banking programme.

Another stumbling block remaining is the aforementioned issue of using conservation bank credits for off-site mitigation in section 7 consultations in federal permits or actions. The Services, federal action agencies, and applicants have been slow to use fish bank credits as conservation measures and have instead focused on on-site mitigation. With so much emphasis (and cost) on on-site mitigation, applicants and permitting agencies are reluctant to add off-site conservation measures because of cost and other uncertainties. One uncertainty is determining effects of a development project on listed fish species and how much mitigation may be required from a conservation bank.[5] The Services have also found it difficult to prescribe the purchase of conservation bank credits as conservation measures as their authority to prescribe conservation measures is limited under section 7 of the ESA. The solution to the problem is to get parties in the consultation process to agree to employ fish banks as conservation measures in the applicant's proposed project.

There has also been a general reluctance to enlist the private sector in endangered species mitigation and recovery because of the assumed higher costs involved. There is an inherent bias against giving mitigation and recovery dollars to 'for profit' corporations despite nearly all recovery plans and regional habitat restoration plans calling for involvement of the 'private sector'. HCP and other public sector in-lieu fee or mitigation programmes have had mixed success in meeting mitigation and recovery goals. Private- sector fish banks entitled by USFWS in recent years have so far proven successful and cost-effective.

USACE recognized early that fish banks have multiple benefits for their own projects in the Central Valley. The USACE Sacramento Riverbank Bank Protection Project (SRBPP) as early as 1997 recognized the value of mitigation banks when it purchased fish credits from the Kimball Island bank (Neff, 2000). With multiple small levee repair projects USACE was having difficulty finding suitable mitigation for each project as they were funded. With projects funded individually it was not possible to develop mitigation ahead of time so credits were available when needed.

Greg Hucks, SRBPP project manager, explains: 'By purchasing the mitigation credit we were able to reduce both the project and schedule cost by eliminating mitigation plan development, lengthy multi-agency reviews, finding and acquiring land, and a host of other time-consuming steps associated with a traditional mitigation process ... the Sacramento River Bank Project stayed on track and helped solve the Reclamation Board's land acquisition dilemma.'

Despite this glowing endorsement the Kimball Island bank remains the only USACE fish mitigation bank in California nearly a decade later. Two uncertainties have hindered further entitlements: (1) market demand for mitigation credits; and (2) whether fish banks produce superior habitat that effectively replaces habitat and habitat functions being lost to development. The first uncertainty relates to mitigation having been focused on on-site or on project-specific off-site mitigation. The second relates to the adequacy of off-site mitigation. Both uncertainties are also related to a belief that banks encourage development by providing readily available mitigation. These problems stem from a general lack of stakeholder understanding and involvement. Umbrella fish bank programmes should go a long way to solving these problems by linking banks to recovery programmes with their built-in stakeholder processes.

Performance of the early banks will also be key to alleviating these concerns. At the Kimball Island bank, five years of monitoring, recently completed, show that all performance measures have been met and that target native fish species use the restored habitats. A review of the sales log for the bank reveals that most of the credits sold were for small-footprint development projects in the Upper San Francisco Estuary. Whether the bank credits adequately mitigate for these small developments has not been in question, rather the general consensus has been that the habitat restored on the larger footprint of the bank site is of greater value than many small 'mitigations' that may have been accomplished if bank credits had not been available. Ultimately, the overall benefits of fish banks will be judged on the basis of their contribution to species recovery, but for now judgement will be based on the quality and quantity, as well as timeliness of habitat developed.

The future of fish conservation banks

NMFS South-west Region has thus far entitled only one fish conservation bank, the Wildlands' Fremont Landing salmon conservation bank, under the Central Valley salmon umbrella conservation bank. The hope is that the bank will help expedite ESA section 7 processes involving NMFS associated with recently funded projects of the Sacramento River Flood Control Project, and assist federal, state and local agencies with their obligation to help recover threatened and endangered fish species in the Central Valley. The Fremont Landing site was chosen because it is one of the last remaining 'natural' riverbank habitats in the leveed reach of the lower Sacramento River with significant opportunity to preserve and restore fish habitat. It is also strategically located at the mouths of three tributary salmon streams.

Wildlands is also working with NMFS North-west Region and the USFWS North-west Region to entitle a conservation bank at a site in the Snohomish River

Estuary of the Puget Sound in Washington State, where development has reduced the capacity of the estuary to sustain populations of listed salmon and bull trout. The site was identified as one of the top priority restoration opportunities in the Puget Sound salmon recovery plan.

Wildlands is also presently in the early stages of developing a salmon and delta smelt conservation bank with the NMFS South-west Region and USFWS Sacramento Office in the upper San Francisco Estuary. Wildlands consulted with NMFS and USFWS on where a salmon-smelt bank should be located within the designated critical habitat of the two species in the Upper San Francisco Estuary. The location chosen was Liberty Island, a known spawning ground of delta smelt and juvenile salmon rearing area. The site is also part of a future planned federal/state wildlife and native fish refuge. The proposed conservation bank will contribute to recovery by acquiring remaining private lands and restoring habitats per the long-term restoration plan for the area.

The Wildlands fish banks and two Oregon ODOT fish banks are the only fish banks entitled or under development at the time of writing this chapter. However, there are many opportunities for fish banking in the Pacific North-west and California because future developments in high-growth regions overlap with many listed species and their designated critical habitats. The Puget Sound area, the lower Columbia River–Willamette River corridors, and California's Central Valley are such high-growth areas. Approximately a quarter of the 149 federally listed fish species are found in these three regions. Other areas of the US with listed fish species and demands for mitigation include the Great Basin, Mississippi–Missouri–Ohio river system in the Midwest, the Great Lakes region, Appalachian Mountain watersheds, the eastern seaboard and the Gulf coast. Alaska may be the only region without a need for fish banks, as it lacks listed species.

Summary

Fish banks are an ideal means of dealing with off-site mitigation needs of development projects that have difficulty in mitigating listed-fish impacts on-site. The key to their value and use is employing banks as conservation measures in the project design during consultations with the Services. Fish habitat takes time to develop and mature and is most valuable in large units or substantial corridors, and thus lends itself to the conservation bank process wherein habitats are preserved and restored in large measure before development impacts occur. Fish banks also can be built with a larger geographic footprint that is more functional than piecemeal small mitigation projects (a general advantage of all types of conservation banks).

Umbrella fish conservation banks fit in well with listed-fish recovery plans because recovery plans focus on ESUs that are generally geographically based. Individual bank sites can then be developed to provide the maximum benefit to the target species population. UBAs fit in well with fish species recovery planning because they allow for strategic planning and standard procedures and rules of engagement in the bank development process, as well as stakeholder involvement.

Low market demand and uncertainties as to the viability of fish banks have limited development of fish banks. The future of fish mitigation/conservation banks depends on whether banks: (1) are politically viable mitigation, (2) contribute significantly to recovery programmes, (3) are cost-effective, and (4) are encouraged, prescribed and employed as conservation measures by the Services.

Notes

1 Sacramento River winter-run chinook salmon (*Oncorhynchus tshawytscha*) – endangered.
Central Valley spring-run chinook salmon (*O. tshawytscha*) – threatened.
Central Valley fall-/late fall-run chinook salmon(*O. tshawytscha)* – candidate.
Central Valley steelhead (*O. mykiss*) – threatened.
2 There are multiple advantages to privately funded restoration. Public funding often depends on government budgets and voter support. The timing of government funding can be precarious to a restoration programme. Public funds are often difficult to encumber and come with many types of constraints, such as in the purchase of lands; whereas private funds have fewer constraints. Private investment can also be used to leverage other funding sources.
3 CALFED is a federal–state programme that includes restoration of fish populations and their habitats in the Central Valley through taxpayer or state-bond funded mitigation for large water supply and flood control projects such as Central Valley Project of the US Bureau of Reclamation and the State Water Project.
4 Fish and fish habitat impacts are more likely a consequence of public-funded development projects (i.e. flood control, water supply and other public works projects), whereas non-fish terrestrial development projects are more likely to involve private-funded development projects. The private-funded projects generally have a greater sense of urgency and less interest in self-mitigating, and thus are more likely to use mitigation banks.
5 Various empirical and mathematical models have been used by USACE and the Services to determine the value of habitat lost at impact sites and the value of habitat conserved or restored at mitigation sites. In California's Central Valley a method developed by USACE and local stakeholders is the standard assessment methodology or SAM model. In Oregon methods developed include the salmon assessment method (again SAM), the wetland accounting method or WAM, and the habitat accounting method or HAM.

References

Bailey, L. L. and Boshard, M. L. (2006) 'Follow the money', in Lackey, R. T., Lach, D. H. and Duncan, S. L. (eds) *Salmon 2100: The Future of Wild Pacific Salmon*, American Fisheries Society, Bethesda, MD, pp99–124

Federal Register (1995) *Federal Guidance for the Establishment, Use and Operation of Mitigation Banks*, vol 60, no 228, pp58605–58614

Lackey, R. T., Lach, D. H. and Duncan S. L. (eds) *Salmon 2100: The Future of Wild Pacific Salmon*, American Fisheries Society, Bethesda, MD

McElhany, P., Ruckelshaus, M. H., Ford, M. J., Wainwright, T. C. and Bjorkstedt, E. P. (2000) 'Viable salmonid populations and the recovery of evolutionarily significant units', *NOAA Technical Memorandum* NMFS-NWFSC-42, US Department of Commerce

Neff, C. (2000) 'Mitigation banks replace lost wetlands', *Engineer Update*, vol 24, no 26, www.hq.usace.army.mil/cepa/pubs/jun00/story8.htm accessed in March 2007

Rahr, G. and Augerot, X. (2006) 'A proactive sanctuary strategy to anchor and restore high-priority wild salmon ecosystems', in Lackey, R. T., Lach, D. H. and Duncan, S. L. (eds) *Salmon 2100: The Future of Wild Pacific Salmon*, American Fisheries Society, Bethesda, MD, pp465–490

US Fish and Wildlife Service (2003) 'Guidance for the establishment, use, and operation of conservation banks', www.fws.gov/sacramento/es/documents/fws_cons_bnk_guide.pdf, accessed in March 2007

11

Getting Two for One: Opportunities and Challenges in Credit Stacking

Jessica Fox

'I'll give you $20,000 for an acre of California Gnatcatcher habitat, but I want the carbon credits for free.' With wetland banking, conservation banking, water quality trading and now carbon markets, opportunities to engage in natural resource markets continue to grow. Investors are attracted by the multiple markets and associated revenue streams, and developers are looking for increasingly comprehensive mitigation. The possibility of credit stacking, selling more than one natural resource credit type on a single acre of land, has emerged with particular interest.

The drive to maximize land management investments will push towards getting credit for all the attributes on a parcel of property. Understandably, landowners will argue that they deserve credit for every ecological benefit they commit to protecting, including the species habitat, sequestered carbon and downstream water quality improvements. On the flip side, if stacking is appropriately regulated, it may create demand for more specific credit types and result in more complete mitigation. With the logistics of markets for ecosystem services still being solidified, the debate about credit stacking is active. This chapter will discuss the opportunities that may exist, the challenges to realizing these opportunities and the potential impact on the environment. Let's start with a hypothetical example.

In 1990, Mr Goody, a wealthy landowner in south-western US, decided to restore a large plot of land that was originally forested. Having previously been clear-cut for its timber, the site was void of ecological and social value. It did not provide habitat, water filtration benefits, carbon sequestration value or recreational value. Rather than opting for a fast growing species, Mr Goody reforested the site with its native moderate growing tree. Seventeen years later in 2007, the site had respectable vegetative growth and was offering both ecological and recreational value.

The area happened to be downstream of a large agriculture operation and now acts as a buffer protecting water quality in the watershed. The nutrient and sediment loading to the nearby waterway, which was previously federally listed as impaired due to the agriculture runoff, has improved significantly. The site is also being colonized by a suite of biodiversity, including a federally endangered bird and a protected salamander.

With the site being restored and its ecological value continuing to increase as it matures, Mr Goody decided to recoup some of his investment by selling ecological credits for the property. After reviewing the market types and talking with several consultants, he estimated that he could get carbon credits, water quality credits and species credits. He first works with Chicago Climate Exchange to establish credits for the carbon rights. Then he works with US Army Corps of Engineers to establish water quality credits for the reduction in nutrient loading, using a baseline for credit generation of 1990 prior to the forest being restored. Finally he works with US Fish and Wildlife Service (USFWS) to establish a conservation bank for the increasing populations of bird and salamanders, and their associated habitat. He sells the various credit types under three different frameworks, the carbon credits were sold within Chicago Climate Exchange's membership, the water quality credits were traded as part of a recently established water quality trading programme in the area, and the species credits were sold with help from USFWS to a developer. Mr Goody more than recouped his investment to restore the site.

Fundamental questions can be asked regarding this scenario. Is it appropriate to be awarded credits under independent programmes (water quality trading, carbon trading, conservation banking) for the exact same land management effort? Is it appropriate for a bank owner to go to different agency offices and obtain agreements for multiple credit types without integration of the agreements? Is there a combination of credit types that are more appropriately stacked than others? And most importantly, do the sold credits represent the necessary project offsets, or is Mr Goody selling the same natural resource value multiple times? Is he double-dipping? It will take several years of discussions to answer these questions. This chapter will lay a foundation for the emerging debate.

Defining terms

Many terms have been used in discussions on credit stacking, including stacking, double-dipping, bundling, multi-use and co-benefits. The issues have been confused by the interchangeable application of these terms, most of whose definitions have not been clearly articulated. This section offers definitions and highlights differences across terms.

Stacking: Acquiring credits under multiple market-based strategies on a single acre of property, where all credits can be sold independently. For example, managing a single acre for conservation banking credits, carbon credits, wetland credits and water quality credits.

Double-dipping: Similar to stacking except that credits double up on the natural resource benefits. When credits are purchased, the necessary mitigation is not achieved because those same ecological values were used up under previous credit sales.

Bundling/Unbundling: How various natural resource values are represented together under one definable unit, or separated out under multiple definable units. For example, a bundled conservation banking credit may include the habitat, sequestered carbon, associated water quality improvements and recreational value. Unbundling will separate out these values into discrete units/credits.

Multi-use: Using a property for multiple compatible uses, with the primary purpose typically protected by a conservation easement or other agreement with a federal or state agency. For example, a conservation bank (with underlying conservation easement), that is used to graze cattle. 'Credits' are not usually sold representing the multi-use.

Co-benefit: The ecological and economic benefits that a bank provides beyond its primary purpose. For example, a conservation bank may provide habitat for non-listed, co-occurring species. Or a wetland mitigation bank may offer positive public relations benefits (and other business perks) for the corporate bank sponsor.

The first two terms, 'stacking' and 'double-dipping', sit squarely at the centre of this chapter. 'Bundling' refers to the legal backdrop that enables stacking to occur, and will be discussed below. 'Multi-use' and 'co-benefit' are not directly related to the debate on credit stacking.

The foundation to stacking is the unbundling

The basis for credit stacking comes from the ability to unbundle natural resource values. Unbundling allows a landowner to claim credit for the nutrient retention in the soils, the wetlands on the soil surface, the upland habitat, the animals themselves and sequestered carbon in the trees, among other things. There is an incentive to unbundle natural resource values to align with their market potential, which now includes species, wetlands, carbon and water quality improvements. The more market types that emerge, the further the unbundling will go.

The bundle of rights concept is commonly taught in US first-year law school to explain how a property can simultaneously be 'owned' by multiple parties. In the bundle of rights theory, ownership of property is compared to a bundle of sticks. Each stick represents a distinct and separate right, which may be the right to use the real estate, to sell it, to lease it, to enter it, to give it away, or to choose to exercise more than one or none of these rights. The rights in the bundle, subject to government limitations and private restrictions, can be sold, leased, transferred or otherwise disposed of individually. For example, one property owner could sell or lease minerals rights and still retain the rights to use the surface area. Another could lease surface rights to one party and lease subsurface rights to another. Still another could sell or lease air rights for aviation, and retain the right to build a house on the land. Thus, the ownership of certain rights may be severed from the ownership of the rest of a property by their being sold, leased or given as a gift to other parties.

Ecosystems are bundles of intertwining values and functions. Natural resource credit markets are attempting to unbundle these functions into marketable units. We have taken the first step by defining the markets: carbon, species, wetlands, etc. But within these markets, credits are still packages of resource values. For example, conservation banking credits include the endangered species, non-threatened species, carbon sequestration in the soil and vegetation, water filtration functions of the land, among others (Figure 11.1). While the credits are sold for the endangered species, they in fact also protect many other ecosystem features which are bundled under the credits on a particular site. Mitigation requirements also represent bundles of natural

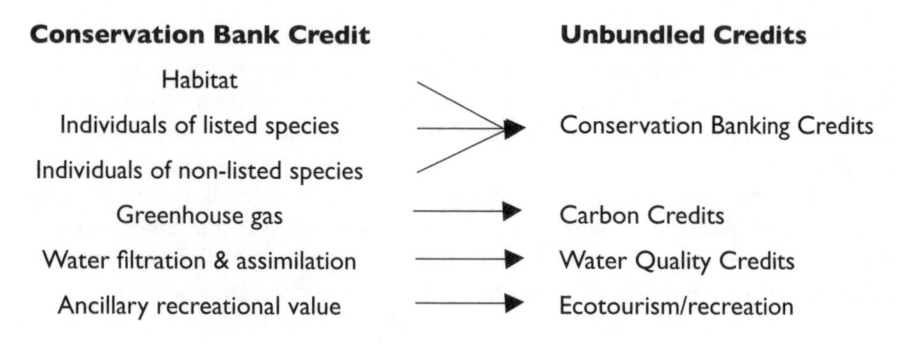

Figure 11.1 *Unbundling a conservation banking credit*

resource values. For example, a developer may be required to mitigate impacts to ten acres of wetlands, but in fact there are many other damaged natural resources that are 'bundled' within the ten-acre requirement, which may or may not be offset. As additional markets emerge, other sticks in the bundle may include ecosystems services such as pollination, agriculture pest control and disease control.

The foundation of credit stacking lies in the accurate unbundling of inherently intertwined ecosystem functions.

Stacking verses double-dipping

Mr Johnson owns a long-leaf pine forest supporting the federally listed red-cockaded woodpecker. Working with USFWS, he establishes species credits for the woodpeckers via a conservation bank agreement (CBA). He sells all of the woodpecker credits, which obligates him to a particular management plan for maintaining the forest. He then plants additional trees, exceeding his habitat maintenance requirements set forth in the CBA. Now working with Chicago Climate Exchange, he generates carbon credits for the planted trees and sells them on Chicago Climate Exchange. In ten years, woodpecker populations increase due to the improved habitat resulting from the planted trees. USFWS awards Mr Johnson additional woodpecker credits to reflect the increased populations. Mr Johnson is now selling both species and carbon credits on the same acres. There is no coordination between Chicago Climate Exchange and USFWS.

In this example, Mr Johnson was probably not double-dipping when he generated his carbon credits. These credits were established based on land management activities that exceeded those required in the CBA. Because additional land management activities were implemented to generate the carbon credits, he could sell all of the carbon credits and all of the woodpecker credits without double-dipping. He did not double sell the carbon values that were bundled with the red-cockaded woodpecker credits. However, when Mr Johnson acquired additional woodpecker credits that resulted from the planted trees (that were already used to generate carbon credits), he risked double-selling. The issue here is the fact that the woodpecker credits are bundled with carbon sequestration values, so separately crediting those same carbon values will result in double counting.

Or consider a conservation bank established for an obligate riparian amphibian. Under the banking agreement, certain activities must be conducted to improve water quality for sediments and nutrients. Being a knowledgeable biologist, the bank owner approaches the Environmental Protection Agency (EPA) prior to making the necessary ecosystem improvements, and establishes a baseline for water quality credits. She implements the necessary land modifications to reduce sedimentation and nutrient runoff, as agreed in her CBA with USFWS. She subsequently returns to EPA to collect credits for the improvements in downstream water quality, which flows into the Chesapeake Bay where a water quality trading programme is in place. EPA reviews her baseline and confirms that the water quality was improved due to her upland management project. She is awarded nutrient credits from EPA for the improvement in water quality. She now sells both water quality credits and species credits acquired by the same land management effort.

In this case, double-dipping is clearly an issue. The bank owner is committed to a particular land management activity under her CBA. Her species credits are bundled together with water quality improvements. Therefore, a credit for the amphibian includes both habitat and improved water quality. Establishing water quality credits separately on the same acres with no additional land management activities will certainly result in double selling of the improved water quality values.

It is possible to stack credits on a single acre without double-dipping, but only if the natural resource values are carefully accounted for. If unbundling is accurate, each credit will be a discreet unit of natural resource value; for example, only woodpecker habitat disassociated from the carbon sequestration value. A separate credit, the carbon credit, will represent the carbon sequestration value. Up until the point of credit sales, a bank owner can decide what mix and credit types he will offer on his property, and then track and account for all sales to avoid double-dipping. Double-dipping occurs at the time of credit sale if the natural resource values were not adequately accounted for and are now being double sold.

The drive to stack credits comes from a financial motivation to maximize returns on land management activities. Presumably, as more credits are awarded per acre, the higher the potential financial return. The move towards stacking and increased returns can be imagined as a continuum (Figure 11.2). On the left side of the continuum there is a bank with a single credit type, the San Joaquin kit fox. Moving to the right, there is a bank with two types of species credits, the San Joaquin kit fox and western burrowing owl. Then there are the banks that offer two different credit types, starting with wetlands and then species – say for vernal pool fairy shrimp. These two credit types can be generated under the same banking agreement and with one agency team, the Mitigation Bank Review Team (MBRT). There are many examples of these types of banks. Next comes what most people consider the official stacking: banks selling two credit types which require multiple regulatory agreements, for example carbon and species, water quality and wetlands, or species and water quality.

As you move from the left to the right, banks can charge more for credits corresponding to their higher habitat value. (A similar continuum can be imagined within each market type. For example, carbon can start with credits based on timber rotation practices, all the way to 'charismatic' carbon which supports biodiversity.)

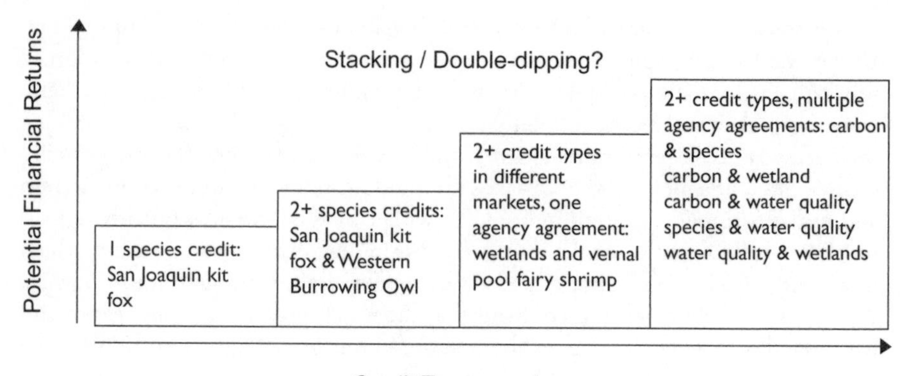

Figure 11.2 *Financial maximization*

The point where double-dipping begins depends on a number of factors, including the accounting system used to award and debit credits, the protocol of unbundling and defining the natural resources on the site, and the specifics of the impacts for which the credits offset. The highest risk of double-dipping occurs when multiple different credits types are awarded on a single acre of property under non-coordinating agency agreements.

The crux of the issue

Property owners can offer a mix of credit types on each acre of property and then optimize decisions about which credits are sold based on current market forces of supply and demand. Hypothetically, this can be accomplished without a negative ecological impact if credit accounting avoids double-sales of the same resource. In fact, stacking may lead to increased scrutiny of the ecological footprint of impacts and result in more detailed and complete mitigation requirements. Therefore, one can argue that there is not a problem in the actual stacking, but in the credit sales. The debate is couched in the details of when double-dipping begins. Do the credit sales represent actual mitigation for impacts, or are natural resource values being double sold?

Avoiding double-dipping depends on whether we can accurately unbundle natural resources values, or mitigate for the errors in our calculations. Trying to separate and neatly define inherently interconnected natural resource values raises fundamental concerns. Given the uncertainties regarding the functioning of ecosystems, it is unlikely that we could accurately account for and correctly define the various attributes of natural communities. Is it possible, for example, to account for carbon credits for redwood trees separately from the biodiversity credits for the animals that live in those trees? Can we separately define the improvement in water quality downstream of the forest from the forest itself? Removing the trees would result in the loss of biodiversity and decrease in water quality – the elements are inextricably linked.

Attempts to parse out various ecological values will most certainly result in miscalculation. While we cannot eliminate the likelihood of errors, there are strategies for limiting their impact. The use of credit ratios is a common approach to mitigate the ecological uncertainties associated with the natural resource banking. Assigning generous credit ratios that account for the large errors in the unbundling of natural resources values may be one option for ensuring that ecosystems are protected.

Another solution may be offered in 'additionality'. Additionality is the approach that credits are awarded only for those land management activities that occur above and beyond previous commitments. Credits can only be generated for 'additional' improvements that would not have otherwise been achieved without proactive measures. This approach is currently being applied to calculate both carbon and water quality credits. Carbon and water quality credits are only approved for management activities that are in addition to what is required to comply with existing permit limits. Applying additionality to a stacking system would mean that landowners cannot get credit under multiple markets for the exact same land management effort. In the opening example, Mr Goody would not be able to acquire multiple credit types unless additional land management activities were required for each credit. Applying this approach sidesteps the need to unbundle natural resource values. In an additionality framework, the credits are still bundles of natural resource values.

In reality, completely apart from credit stacking itself, there are few rigorous assessments of the ecological impact of any of the market-based approaches. Discussions are active regarding the functioning of the wetlands in wetland banking, the quality of habitat in conservation banks, the optimal methods for greenhouse gas accounting in carbon markets, and the real improvements resulting from water quality trading programmes. Further, the permanence of these efforts, how long the ecological benefits of these banks are maintained, is unsubstantiated. The answers to these more fundamental questions need to be resolved before we have any concept of the ecological impact that credit stacking will have apart from banking itself. These realities put the credit stacking debate in the appropriate ecological perspective.

Business and regulatory hurdles

The potential benefits of credit stacking include increased financial returns on a single parcel of property, variability of revenue streams for the different credit types that can be optimized in real time, one-stop shopping for mitigation seekers and, possibly, environmental advantages due to more comprehensive mitigation. It is the financial opportunities associated with credit stacking that are the most compelling. However, even apart from the core ecological challenge of unbundling discussed above, several issues lie in the way of realizing these opportunities. These issues can be organized by their business and regulatory elements.

Business issues

There is an assumption that credit stacking is preferable from the bank owner's perspective, as it offers more opportunities to receive financial returns for a single acre of land. However, stacking may lead towards more stratification in dollar allocation for natural resources. For example, consider a bank that is selling San Joaquin kit fox credits for $75,000. If the bank owner decides to establish carbon credits on the site, she may simply transfer the greenhouse gas sequestration value from the currently bundled kit fox credit to the new carbon credit. Now she has kit fox credits worth $50,000, and the carbon credits are worth $25,000. Because she reduced the ecological value of the kit fox credits by subtracting the carbon value, the price went down. Due to the stratification in credit values, she really hasn't increased the financial potential on the property. Rather, she may have made her life more complicated by parsing out previously bundled credit values, now having to market her credits to many more types of buyer. If multiple credits are awarded at one property, the bank owner may need to do an optimization analysis to decide how many of which credits to establish.

Of course, there will be no drive to stack credits unless there is demand for the various credit types. And there will be limited demand for credits unless there are strong regulations requiring mitigation, or the threat of them (as in the case of climate change). Unbundling natural resource values in order to generate credits obligates one to unbundle the various values on the impact side, and visa versa. Currently, if a wetland is impacted, a developer must compensate for the acres of wetland impacted and any specific endangered or threatened species on the property. In a system that is unbundled, they may have to account for wetlands, carbon credits and water quality impacts separately. One by-product of stacking will be much greater scrutiny on unbundling the ecological footprints of impacts of businesses and government, which will lead to more detailed mitigation requirements.

It may be that once the two sides of the equation are equalized – with the full impacts accounted for and all the mitigation credited – the current system is preferred by both banker and developer. But even if unbundling is not advantageous for the buyer or seller, from an ecological standpoint it is important to fully identify and quantify the functions that are lost even if they aren't bought and sold separately.

Regulatory issues

Even if the ecological and economic perspectives still argue for credit stacking, there are logistical challenges at a regulatory level. The success of the current ecosystem markets rely on regulatory approval, oversight and enforcement. Coordination between different agencies is necessary to approve credits and ensure that the selling of credits represents the intended mitigation. Typically, the US Army Corps of Engineers takes the lead on approving and monitoring wetland banks, USFWS approves and monitors conservation banks (plus state departments of fish and game when state-listed species are involved), and EPA oversees water quality trading (authority is passed to the states in many cases). Because carbon credits are being established and traded on a purely voluntary basis at this time, there is no official regulatory oversight in the US. However, private organizations, such as Chicago Climate Exchange in the US, are acting in an oversight role to approve credits and transactions.

Given the various groups involved in the different credit types, the obvious concern is how the agencies will coordinate with each other to oversee and monitor sites that have multiple credit types. There is some track record here. For properties

Table 11.1 *Agencies involved in different credit types*

Credit Type	Primary Oversight: Agency Office / Division
Carbon	Private organizations. Eventually may be DOE and/or EPA.
Endangered Species	FWS, Endangered Species Program. Some state Departments of Fish and Game.
Wetlands	US Army Corps of Engineers.
Water Quality	EPA, Office of Water. Some states.

that support both wetlands and endangered species, the agencies have cooperated to use one banking agreement. However, the time required to approve even a simple bank for one endangered species has ranged from eight months to several years. Banks requiring multi-agency coordination, such as with wetlands and carbon, will probably be significantly more protracted.

Even after the banking agreements are established, the agencies need to ensure that the credits are being accounted for accurately. In a conservation banking system, bank ledgers are submitted to USFWS so they can monitor and account for the number of credits sold. There already are challenges in gathering these accounting ledgers and reviewing them for errors. If it were possible to acquire multiple credit types on a single acre, the accounting task would become even more burdensome. For example, a bank selling credits for both species and carbon would have several different agencies involved and the accounting sheets would have to be clear on whether they sold each acre for species only, carbon only, or species and carbon.

Given the history, the task of establishing relationships across currently uncoordinated agency offices is substantial. Unless it is possible to establish protocols that sidestep the need for agency coordination, it is reasonable to predict issues with tracking and monitoring banks that sell multiple credit types.

Conclusion

There is significant energy being applied to investigate stacking opportunities. Several groups are looking into the possibility of acquiring wetland credits for the wetland itself plus water quality credits for the improvement in downstream water quality. One ambitious effort led by the Wetland Initiative in Chicago is looking at piling on carbon credits into that mix (*Grand Rapids Press*, 1 July 2007). Recently, Standard Carbon LLC was the first to establish carbon credits for the greenhouse gas offsets provided for in a wetland mitigation bank (www.standardcarbon.com). Some existing conservation bank owners are also interested in getting carbon credits. While the specifics of the credit stacking scenarios presented in this chapter are hypothetical, the interest in seeing similar situations realized is strong. And by rewarding landowners for the full ecological benefits that they protect, incentives are created to protect even more resources.

The issue comes when credit sales no longer represent mitigation – when stacking becomes double-dipping. Two possibilities for avoiding double-dipping are to adequately mitigate the inevitable errors in unbundling natural resource values, or to apply the approach of additionality. While credit ratios may help mitigate ecological uncertainties in unbundling, even this may not be enough to withstand ecological scrutiny. The safest approach may be to use the criteria of additionality; however, more analysis is necessary before a strong recommendation can be offered.

Ultimately, economic drivers to maximize investments will push towards a resolution of the various regulatory, business and ecological challenges around stacking. The bottom line is that there are opportunities, there are challenges to realizing these opportunities, and the value to the environment is not yet clear. Soon a regulatory decision will be forced clarifying what is allowed and what isn't. In the meantime, Mr Goody and ten others like him are quickly emerging from these pages to offer the most comprehensive and profitable mitigation ever.

Reference

Grand Rapid Press 'Restored wetlands would filter air; group seeks permit to convert Illinois farmland', Michigan, 1 July 2007, www6.lexisnexis.com/publisher/EndUser?Action=UserDisplayFullDocument&orgId=1925&topicId=100002047&docId=l:638470488&start=10

12

The Marine Leap: Conservation Banking and the Brave New World

Tundi Agardy

Introduction

The considerable strides taken towards developing conservation banking (described in previous chapters) have clearly paved the way for private sector investment in the conservation of natural ecosystems. It is difficult to imagine what this brave new world of private sector conservation will look like in 10, 20 or 100 years, but it probably will have dramatic impacts not only on our surroundings on land, but also on the marine environment. What follows is an exploration of some of the most promising new tools for promoting marine conservation that this brave new world will bring forth. Many of these tools are still in the pilot – or even theoretical – stages of their development; they are outlined here in the hope that they will help us peer a bit deeper into the future of market-based strategies for protecting the oceans upon which we all depend.

The Millennium Ecosystem Assessment (MA) has shown that almost 40 per cent of the global population lives within a thin band of coastline representing only 5 per cent of the earth's terrestrial area (MA, 2005). Importantly, coastal populations are growing faster than populations in most other areas and the attendant destruction of habitats is leading to the erosion of ecosystem services that are notoriously difficult to replace. Mangroves and coastal marshes maintain hydrological balances, contribute to freshwater recharge of aquifers, prevent erosion, regulate flooding, and buffer land from storms. Rock and coral reef habitats also buffer land from storms. With some 71 per cent of the world's coastal population living within 50 kilometres of an estuary, 31 per cent living within 50 kilometres of a coral reef system, 45 per cent living within 50 kilometres of mangrove wetlands, and 49 per cent living within 50 kilometres of sea grass ecosystems, the ecosystem services provided by coastal habitats are under more pressure than ever before (MA, 2005).

In the wake of widespread concern for the deteriorating state of our oceans and coasts, new approaches to coastal management and conservation are desperately needed. Fortunately, innovative, market-based mechanisms to manage resource use and protect ecosystems have been tested at small scales with promising results.

Species banking

Salmon

Salmon, emblematic and highly valued, have been the pioneering set of species pushing species banking out into the wet. Their anadromous habits create conditions where individuals or corporations with property rights to riparian tracts of land can trade credits to protect vital habitat. If a proposed development will have a negative impact on salmon habitat (or a related species), the developer (private or public) can purchase credit in a salmon habitat bank (Carroll, 2006). For-profit companies in the US have begun to create such banks by identifying key salmon habitats, purchasing parcels and banking them, and even restoring habitats to improve salmon health (see Chapter 10). Kimball Island mitigation bank, owned by Wildlands Inc., began this innovative trading trend by investing in salmon protection, and other ventures are following in their footsteps. Such banks harness private sector investment in concert with both public and private non-profit conservation initiatives, building over time to capture the most critical habitats and avoid mitigation resulting in habitat fragmentation (Bruggeman, 2006).

Other marine species

So it is that salmon, with their connection to ocean, stream and forest, have brought us terrestrial creatures out into the salty brine. But other marine species are ripe for receiving the benefits of conservation banking, too. The most suitable will probably prove to be those that are indeed marine, yet tied in some way to specific terrestrial habitats such as nesting or feeding areas -- species which at the same time have been flagged as endangered or vulnerable. The recognition that such flagship species are threatened creates both the public interest in searching for innovative ways to protect them, and establishes the regulatory/policy environment favourable to species banking efforts. The tie to terra firma also seems critical in these early stages of species banking forays into the marine realm – for it is only on land that property rights are well understood. The link to specific terrestrial habitats creates business opportunities to protect or even enhance riparian and coastal habitats, thereby potentially generating credits for species protection. Such credits could be generated through enhanced reproductive habitat, creation of habitat shelters, increases in food quantity and quality, and even water quality improvements (J. Fox, pers. comm.). Linking such marine species banking with water quality markets could then leverage investments in marine conservation as a whole, by addressing environmental issues through two different, but related, perspectives.

One group of similarly threatened and valued animals for which conservation banking schemes make sense is sea turtles. Like salmon, these organisms spend the bulk of their lives in the oceans, but they must return to land to reproduce. One of the greatest threats to sea turtle species worldwide, after fisheries-related mortality, is loss of nesting habitat. Wide open sandy beaches are as highly valued by coastal developers and tourists as they are by nesting turtles – and in the race to use these habitats, the sea turtles inevitably lose. While laws are on the books in many countries that attempt to control development impacts on nesting beaches, the reality is that coastal development is proceeding at blinding speed in the tropics, and habitats and their ecosystem services are being degraded as a result. Though in rare cases coastal development and nesting sea turtles can co-exist, the trend in nesting beach degradation and loss spells doom for sea turtles in many parts of the world.

If sea turtle critical habitat were identified, purchased and banked to allow trading, coastal developers could greatly lessen their impact on these charismatic species. More critical habitat would be protected in banks, and trading would allow those developments that avoided impact to nesting habitat or worked to restore nesting habitat to sell credits to coastal developers expecting to negatively impact nesting beaches. In addition, unavoidable sea turtle by-catch could be offset by protection and restoration of nesting habitat (N. Carroll, pers. comm.). Conservation benefits from greater fund flows for protection and restoration, the private sector benefits because they can proceed in a known, low-risk regulatory environment and utilize business principles that they understand, and the public benefits because awareness about sea turtles grows and their protection is much enhanced.

One thing that is obvious from terrestrial species banking initiatives, however, is that a strong regulatory environment is an absolute necessity for these species banks to gain traction and become a force in conservation and development. In the US, the strong Endangered Species Act (EPA) of 1973 allows for significant investment in identification of critical habitat, and for restrictions to be placed on both public and private activities that affect listed species. Species banking has become a very popular tool in places where development is highly constrained by the presence of many listed species, as is the case in California. It is catalysed in situations where a strong regulatory environment creates incentives for conservation and disincentives for unsustainable development (i.e. mitigation requirements are strictly enforced), where the business environment is secure and risks are perceived to be few, and where opportunities to beat the system by cheating are rare. Policy and regulatory institutions are already creating these enabling conditions, even if not explicitly for the purpose of setting the stage for conservation banking. Nonetheless, the conditions improve daily, and as they do, the fear of risk in the private sector decreases.

By-catch markets

Focusing on species and their value in providing ecosystem services can take us beyond 'conventional' marine species banks (everything is relative, after all) to some truly unconventional ideas that are suddenly gaining traction. Ecotrust, for instance, is exploring the idea of establishing markets for the trade of by-catch credits in commercial fisheries. Everyone realizes that by-catch, or the incidental capture and waste of non-target species in commercial fisheries, has enormous impacts on marine ecosystems and presents huge environmental, and in some cases economic, costs (see Worm et al, 2006, for example).

The fact is, by-catch reduction strategies have failed in many commercial fisheries, especially those far removed from the seeing eye of coastal authorities, whether out in the Exclusive Economic Zones (the 200 nautical mile limit of most countries' jurisdictions), or on the last frontier of the High Seas. These failures have spurred great interest in finding alternative ways to reduce waste and ecological degradation, while at the same time not crippling the industry with economic inefficiencies. The idea behind a by-catch market is to identify an acceptable level of by-catch, and allow fisheries entrepreneurs to develop ways to reduce by-catch, create by-catch credits and sell them to those unwilling or unable to reduce by-catch in their fishing operations. Because it is in the best economic interest of commercial fishers to do so, this is probably the best hope for curtailing the astounding rates of incidental catch in US and other highly developed fisheries.

Conservation banking and property rights at sea

The momentum generated by these fledgling market-based approaches to coastal and marine conservation suggests there may be a place for marine species banking in the future – but taking species banking off-shore leads us to truly uncharted territory. Perhaps the biggest challenge to species banking in the marine realm is the common property regime of oceans, which makes ownership difficult to establish in ocean habitats (McCay, 2000; Raymond, 2003). Fortunately, the US and the rest of the world seem to be headed towards radical reform in other aspects of ocean and coastal management. A recent article in the journal *Science* points to the value of spatial management in improving ocean governance (Crowder et al, 2006), and two states – California and Massachusetts – are set to develop ocean zoning plans within state waters. The US federal government is looking to these trend-setting state initiatives with interest, and is currently debating the extent to which developing zoning for federal waters would improve management. Outside the US, New Zealand, Belgium, Germany and other countries are all embarking on full ocean zoning for their waters, and other countries such as Tanzania are developing small-scale zoning plans that may well be scaled up to the national level at some point in the future.

Zoning can clarify boundaries of ownership and more fully codifies use rights, which in turn reduces risk for those wishing to engage in coastal resource or ocean-space trading (Agardy, 2007). While not currently on anyone's radar screen (yet), it is also possible that zoning plans could be developed that establish special trading areas for certain Payments for Ecosystem Services markets. Such trading areas would delimit geographical areas where property and use rights are well-established and credits could be generated. Working with coastal and ocean planners in the early stages of development of zoning plans would ensure that zoning plans meet their full potential to engage the private sector in the important task of protecting species and habitats, and maintaining crucial and valuable ecosystem services (Worm et al, 2006).

Conclusions

The future and potentially dramatic paradigm shift that will occur in marine conservation will be witnessed by the leap from generating credits from conservation activities on terrestrial/riparian habitat to generating credits from conservation activities in open water (controlling by catch, reducing pollution, etc.). At present it is easier to generate marine credits for species that are linked to specific terrestrial habitat versus wholly marine species because the terrestrial link means that property rights are defined (simplifying the question of who gets the credits), and because anchoring habitat protections to land as opposed to the more fluid marine environment makes protections easier, more transparent and better tied to regulatory requirements under the EPA. However, with each successful new initiative involving these 'bridge species', we move one step closer to being able to use the full range of market-based tools for conserving ocean ecosystems and the marine life they support.

Given the importance of coastal and marine services, the move to develop new approaches and achieve greater engagement of the private sector provides significant hope for the future. There are almost infinite possibilities for expansion of habitat and species conservation schemes. These initiatives can be thought of as being in one of two categories. Incentive-driven solutions include reducing coastal storm risk, increasing tourism value, increasing marine product value and reducing costs. Then there are regulation-driven solutions, such biodiversity offsets, mitigation banking, cap and trade, and the sorts of marine species banking described in this chapter. The plethora of innovative new solutions is not merely the fantasy of conservationists, however. This train's already left the station, and we're already well on our way to that brave new world.

References

Agardy, T. (2007) 'Ocean zoning is coming! Ocean zoning is coming! Music to some ears; striking fear in the hearts of others', *Ocean Observer*, February, World Ocean Observatory, available at www.thew2o.net/archive_new.html?id=29., accessed February 2007

Bruggeman, D. (2006) 'Market-based approaches for re-connecting the landscape', *Banking on Conservation: Species and Wetland Mitigation Banking*, 36–38, The Ecosystem Marketplace, San Francisco, CA, available at http://ecosystemmarketplace.com/pages/newsletter/4.3.06.html, accessed December 2006

Carroll, N. (2006) 'Conservation banking emerges in the Pacific northwest', *Banking on Conservation: Species and Wetland Mitigation Banking*, 36–38, The Ecosystem Marketplace, San Francisco, CA, available at www.ecosystemmarketplace.com, accessed December 2006

Crowder, L. B., Osherenko, G., Young, O. R., Airame, S., Norse, E. A., Baron, N., Day, J. C., Douvere, F., Ehler, C. N., Halpern, B. S., Langdon, S. J., McLeod, K. L., Ogden, J. C., Peach, R. E., Rosenberg, A. A. and Wilson, J. A. (2006) 'Resolving mismatches in U.S. ocean governance', *Science*, vol 313, pp617–618

McCay, B. (2000) 'Property rights, the commons, and natural resource management', in Kaplowitz, M. D. (ed) *Property Rights, Economics, and the Environment*, JAI Press, Stamford, CT, pp67–82

Millennium Ecosystem Assessment (2005) 'Coastal systems', *Ecosystems and Human Well-Being*, vol 1, ch 19, Island Press, Washington DC, pp513–549

Raymond, L. (2003) *Private Rights in Public Resources*, Resources for the Future, Washington DC

Worm, B., Barbier, E. B., Beaumont, N., Duffy, J. E., Folke, C., Halpern, B. S., Jackson, J. B., Lotze, H. K., Micheli, F., Palumbi, S. R., Sala, E., Selkoe, K. A., Stachowicz, J. J. and Watson, R. (2006) 'Impacts of biodiversity loss on ocean ecosystem services', *Science*, vol 814, pp782–787

Part IV

Going Global

13

Biodiversity Offsets

Kerry ten Kate and Mira Inbar

Introduction

Currently the world is witnessing an unprecedented loss of biodiversity. Some 10–30 per cent of all mammal, bird and amphibian species are threatened with extinction. A major cause of this loss is the destruction of natural habitats by developments in the agriculture, forestry, oil and gas, mining, transport and construction sectors, among others. At the same time, countries rely on these developments for economic growth and for products, services and jobs.

Conservation groups, governments and companies are looking for practical mechanisms that integrate environmental issues into planning and development. Biodiversity offsets are conservation actions intended to compensate for the residual and unavoidable loss of biodiversity incurred during new developments. They are increasingly attracting attention because of their ability to achieve more, better and higher priority conservation and livelihood outcomes than currently occur in the context of infrastructure projects. Biodiversity offsets not only rehabilitate sites but also address the company's full impact on biodiversity at the landscape scale, thus assisting companies to manage their risks, liabilities and costs. In addition, biodiversity offsets can support sustainable livelihoods, addressing some of the underlying causes of biodiversity loss incurred by human use of natural resources.

Conservation banking in the US is but one country's interpretation of how to perform biodiversity offsets, one that requires a particular regulatory framework. Many countries are exploring the concept of biodiversity offsets in a broader sense. This may involve introducing new national policy and legislation to regulate offsets or it may involve developers undertaking offsets on a voluntary basis, since the business case for doing so is compelling.

This chapter will:

- discuss the meaning of biodiversity offsets and the need for them;
- explore recent history and the current context for work on biodiversity offsets;

- describe the two main motivations for biodiversity offsets: regulation and the business case;
- describe some of the challenges inherent in designing best practice biodiversity offsets;
- introduce an international programme to develop practical experience, guidelines, and international and national policies on biodiversity offsets;
- offer some conclusions about future opportunities and next steps for biodiversity offsets.

The meaning of biodiversity offsets

Biodiversity offsets can be defined as 'conservation actions intended to compensate for the residual, unavoidable harm to biodiversity caused by development projects, so as to ensure no net loss or preferably net gain of biodiversity. Before developers contemplate offsets, they should have first sought to avoid and minimise harm to biodiversity' (ten Kate et al, 2006). Biodiversity offsets have a slightly different connotation in international discussions than they do in the US, where the Endangered Species Act (ESA) and the Clean Water Act (CWA) drive mitigation of habitat impacts. While in the US the trigger for offset activities like conservation banking is regulatory compliance, in other parts of the world there are often no comparable regulatory requirements for biodiversity offsets. However, all around the world, developers, financial institutions and conservation groups are looking at the business case for undertaking offsets on a largely voluntary basis.

Thus biodiversity offsets have much in common with the mitigation measures with which those in the US will be familiar, but there are some key distinctions (see Table 13.1).

The scope of activities which constitute a biodiversity offset are currently being debated, but there is emerging consensus that offsets are actions that lead to measurable *in situ* conservation outcomes. Other activities, such as capacity building with policy makers and public employees such as park managers are acknowledged as very important, but often seen as part of a package of activities needed to make the offset viable, rather than being part of a biodiversity offset itself.

Developers should pursue biodiversity offsets only at the end of the mitigation hierarchy, after they have reduced and alleviated residual environmental harm as much as possible. The more significant the biodiversity that is impacted the greater the need for emphasis on avoidance of harm. Biodiversity offsets can be used to compensate for the residual impact to biodiversity that cannot be mitigated on-site and balance the impact of the project.

The need for biodiversity offsets

Standard corporate environmental management approaches, such as the use of environmental impact assessments and environmental management systems, rarely focus on the threats that developments pose to biodiversity. Approaches to

Table 13.1 *Comparison of US mitigation with biodiversity offsets in other parts of the world*

Wetland mitigation and conservation banking in the US	***Policy and emerging international practice on biodiversity offsets worldwide***
Policy goal of 'no net loss' or 'net gain'	In some countries (approximately 30), there is a policy goal of 'no net loss' or 'net gain' in some circumstances (eg for significant impacts on listed species). However, 'no net loss' or 'net gain' of biodiversity is emerging as voluntary best practice, and may also be a requirement for access to finance. See 'business case', below.
Follows mitigation hierarchy of avoid, minimize and only then 'mitigate' (which is considered equivalent to 'offset')	Follows mitigation hierarchy of avoid, minimize and mitigate, then offset. (Outside the US, for instance in Europe, 'mitigation' is often defined as a means of reducing impacts, but may not get as far as compensation for all residual loss. Therefore, 'offset' is regarded as the last step, covering the residual, unavoidable loss after minimization/mitigation.) In some countries (particularly Spanish- and to some degree French-speaking countries), the term 'compensation' is used for 'offset'.
US	International: policy measures specifically on biodiversity offsets in about 30 countries. In addition, voluntary biodiversity offsets are emerging as best practice worldwide, led principally by listed multinational companies.
Motivation: regulatory compliance	Motivation: business benefits, including licence to operate and access to capital.
Covers listed wetlands and listed species	Policy instruments often cover listed species or habitats, or impacts above a certain threshold of gravity. Voluntary biodiversity offsets cover impacts on any components of biodiversity (i.e. any ecosystems, habitats, species) that give rise to business risk and opportunity.
Focuses solely on intrinsic values of biodiversity, through priority listing of particular ecosystems (e.g. wetlands) and species (as listed in the ESA)	Addresses not only intrinsic values of biodiversity, but also people's use values. (In developing countries, the priority biodiversity values are often those associated with people's livelihoods – e.g. medicinal plants, fuelwood, agroforestry species – or with other social and cultural values. In developed countries, biodiversity offsets may address people's amenity values – e.g. recreation.)

Box 13.1 *Types of offset activities*

While appropriate offset activities will vary from site to site, a range of different land (and marine) management interventions are typically involved in biodiversity offsets, including:

- Restoring or rehabilitating degraded areas: replanting indigenous trees on degraded land.
- Strengthening ineffective protected areas: investing in additional management activities in neglected zones of a forest reserve (i.e. replanting degraded areas or removing alien invasive species) and therefore improving its conservation status.
- Protecting threatened areas: averting the risk of development on a piece of land by putting in place conservation management either by working with communities or with government.
- Addressing underlying causes of biodiversity loss: working with communities to develop alternative livelihoods through sustainable land management (i.e. woodlots, cleaner burning stoves, etc.) and stopping unsustainable activities (fuelwood chopping, crop plantation in forests).
- Establishing corridors: identifying and securing the conservation management of land that provides biological corridors between protected areas.
- Establishing buffer zones: for instance, around a national park lacking a buffer zone.
- Zoning marine areas: for example, demarcating and protecting areas important for feeding and breeding. Working with companies and communities to avoid exploitation in these areas. Supporting sustainable aquaculture initiatives for communities to compensate for lost income.
- Securing migration paths: establishing interventions to secure migration paths.
- Removing livestock from a biologically sensitive site that is being overgrazed.

mitigating the environmental impact of new capital projects tend to seek engineering solutions to reduce impacts such as noise and air pollution, and often neglect biodiversity considerations, such as impacts on ecosystem health or the biodiversity use by local communities. Companies may partly rehabilitate project sites, leaving the affected ecosystem degraded to the point where the area has little conservation or biodiversity value. Overall, traditional environmental management does not measure the full impact of development projects on biodiversity and, therefore, does not fully compensate for that impact. Because of this, the legacy of development is one of continued loss of biodiversity and often a residual negative impact on local communities' use and enjoyment of natural resources.

The poor environmental, socio-economic and health legacies of many development projects in the past have harmed the reputation of companies and created financial liabilities for them. Companies are increasingly aware that biodiversity offsets can help them manage biodiversity-related risks and may also open up new business opportunities. Companies see that with judicious biodiversity offsets they can both rehabilitate sites and provide significant and enduring conservation results at the landscape scale for costs comparable to some on-site rehabilitation. In addition, they can use offsets to address local communities' biodiversity-related livelihood priorities, thus tackling a common cause of local biodiversity loss and also securing the social licence to operate, which they prize.

When biodiversity offsets are well-designed and undertaken in appropriate circumstances, they can help planners and developers balance development needs with environmental concerns, by integrating conservation into development planning. By engaging with government and local communities in the design of offsets, companies can focus conservation efforts on landscape-level and regional priorities and help meet national and international biodiversity goals. Offsets designed with the landscape and bioregional context in mind are more likely to offer spatial connectivity, and thus conservation value, than small, isolated offset areas. In addition, there is growing interest worldwide in conservation-based markets (such as mitigation banking in the US), which allow companies to minimize their costs through economies of scale, achieve results efficiently and manage liabilities by outsourcing the actual conservation activities to expert third parties. This is also of interest to conservation groups and government planners since it can contribute to land-use planning and also create aggregated offset areas on a scale that achieves significant conservation outcomes in high biodiversity-value areas.

However, while biodiversity offsets offer these and other potential benefits, they also pose considerable risks both for business and conservation. There is concern that offsets could be one step on the slippery slope to allowing inappropriate developments to go ahead, especially in the few remaining fragile and pristine ecosystems around the world. If undertaken without the appropriate measurement and proper planning for long-term outcomes, offsets could very well fail to deliver their desired conservation and livelihood outcomes. In addition, companies will not undertake offsets if the transaction costs are too high, or if they see offsets as leading them into the role of surrogate governments in places where the authorities apparently do not have the capacity to fulfil their conservation mandate.

Recent history and current context

Currently, a number of countries have regulations in place that mandate environmental compensation for development impacts. Conservation and wetland banking in the US, which have been under development for more than two decades, have offered guidance for legislation in other countries. Australia, Brazil, Canada, the European Union and Switzerland already have legislation that mandates compensation for certain kinds of biodiversity loss. Table 13.2 lists compensation schemes around the world. In addition, other countries have expressed interest in developing enabling frameworks for offsets, including France, South Africa, Madagascar, Mexico and Uganda.

Motivations for biodiversity offsets

There are three principal motivations that drive private and public developers to undertake biodiversity offsets. They include laws requiring offsets, laws that do not explicitly require offsets but facilitate dialogue on the conditions for planning permission, and a related but distinct business case for voluntary offsets.

Laws requiring offsets

Legal provisions in a number of countries require the creation, restoration or *in situ* conservation of habitats to compensate for the damage caused by development activities. Although enforcement of such legal requirements varies, depending on the capacity of the government concerned, this is the principal motivation for companies to undertake biodiversity offsets in countries where such legislation exists.

Table 13.2 *Examples of legal requirements for biodiversity offsets*

Country	*Programme*	*Legislation*	*Policy goal*
US	Species mitigation (of which conservation banking is one tool for mitigation)	ESA 1973, as amended, and 'Guidance on establishment, use and operations of conservation banks'	to offset adverse impacts to threatened and endangered species
	Wetland mitigation	CWA 1972 Chapter 404(b)(1) and the US Army Corps of Engineers regulations (33 CFR 320.4(r))	'no overall loss of values and functions' (1990); 'net gain' (2004)
Australia	New South Wales	'Green offsets for sustainable development', Concept Paper (2002); Native Vegetation Act (2003) & subsequent regulations (2005); The Threatened Species Conservation Amendment (Biodiversity Banking) Bill (2006)	'net environmental gain'

Country	***Programme***	***Legislation***	***Policy goal***
	Victoria	Native Vegetation Management Framework- (2002) & subsequent amendments to related Acts; Bush Broker – native vegetation credit registration & trading, Information Paper (2006)	'a reversal, across the entire landscape, of the long-term decline in extent and quality of native vegetation, leading to a net gain'
	Western Australia	Native Vegetation Act (2003); Environmental offsets, Position Statement No 9 (2006)	'net environmental benefit'
Brazil	Forest regulation and national system of conservation units	Lei No. 4771 of 1965; Lei No. 14.247 of 22/7/2002, Lei No 9.985 of 18/7/2000, Decreto No 4.340 of 22/8/2002	no net loss of habitat under a defined minimum forest cover for private landholdings
Canada	Fisheries Act	R.S. 1985, c. F-14, Policy for the management of fish habitat (1986), and 'Habitat conservation and protection guidelines, 2nd edn (1998) (see especially subchapter 35(l) and subchapter 35(2) of the Fisheries Act)	no net loss in capacity of habitat to produce fish
European Union	Habitats and birds directive	Council Directive 92/43/ EEC of 21 May 1992 on the conservation of natural habitats and of wild fauna and flora and Council Directive 79/409/EEC	maintain overall (ecological) coherence of the sites

Source: ten Kate et al (2004); McKenney (2005)

Environmental impact assessments and planning law

In many countries, Environmental Impact Assessments (EIAs) provide the necessary framework for governments to negotiate biodiversity offsets with developers. From a company's perspective, an operation's site environmental management plan is generally linked to issues that arose during the EIA. However, in order for the EIA to act as a trigger for biodiversity offsets, the EIA system itself needs to be robust and transparent, so that the full mitigation hierarchy is followed, biodiversity offset negotiations take place and offsets are not seen as attempts to 'buy off' officials.

In addition, EIAs are often conducted on a timescale that frequently does not synchronize with the biodiversity being studied. For instance, it may take more than a year to understand potential seasonal impacts and to consider which aspects of a site's biodiversity are priorities for conservation efforts. By contrast, EIAs are often completed within a period of six to nine months. Finally, some conservation organizations have expressed concerns that, since EIAs are usually paid for and approved by the companies causing the environmental damage, they may underestimate the damage caused or the offsets needed to compensate for said damage.

In many countries, the planning process, with its formal system of applications and enquiries, offers another potential trigger for dialogue on biodiversity offsets between developers and regulators. Indeed, environmental and social works are often required as a condition for planning approval, or as a form of 'planning gain'. For instance, in the UK, section 106 of the Town and Country Planning Act has often been used by authorities to require developers to undertake compensatory conservation activities. Just as with EIAs, certain underlying conditions may be needed for this trigger to work, such as fiscal tax breaks and density bonuses.

The business case for voluntary biodiversity offsets

Some companies have sophisticated approaches to mitigating loss of biodiversity and rehabilitating former operating sites. A growing number of companies, especially in the extractive sectors, have demonstrated that there is a strong business case for going beyond mitigation to compensate for the full impact that their developments have on biodiversity. Companies are increasingly looking to demonstrate good practice on environmental issues to secure their licence to operate and access to capital, to obtain permits rapidly and operate cost-effectively, and to maintain a competitive

Box 13.2 *The business case for biodiversity offsets*

- Ensure continued access to land and capital. For certain industry sectors, particularly mining and oil and gas, access to land and sea is a principal driver of financial success. (See, for example, Goldman Sachs [2004, 2005].) Those companies that can demonstrate best practice on the management of biodiversity (including biodiversity offsets) can achieve faster permit and concession negotiations that produce earlier revenues and considerable savings.

- Competitive advantage of favoured status as a partner. Companies often compete against each other for extractive and utility concessions or to lead or take part in consortia involved in major extractive operations. Companies that are not able to demonstrate best practice in management of all key environmental issues, one of which might be biodiversity, could be at a competitive disadvantage relative to others in securing concessions (Grigg and ten Kate, 2004).
- Increase investor confidence and loyalty. Investors and creditors are increasingly examining whether the companies in which they are invested can assess and manage both financial risk and what is sometimes known as 'non financial' or 'GSEE' risks: risks related to governance, social, environmental and ethical issues. Mainstream investors such as Goldman Sachs, Citigroup, UBS, Schroders, Insight Investment and F&C Asset Management investigate the quality of companies' risk management on these topics, of which biodiversity is just one. (See, for example, reports by Insight Investment (Grigg and ten Kate, 2004; Foxall et al, 2006) and F&C Asset Management, 2005.
- Reduce risks and liabilities. Companies that do not manage biodiversity effectively are exposed to potential business risks: liabilities, damage to reputation and increased operating costs, as experiences in the mining and oil and gas sectors over the last two decades have shown (e.g. Miranda et al, 2003).
- Undertake projects that might not otherwise be possible, because residual impacts that were not offset would be considered unacceptable.
- Ensure strong and supportive relationships with local communities, government regulators, environmental groups and other important stakeholders. Examples in the mining and oil and gas sectors reveal project costs of hundreds of millions or even several billion dollars stemming from delays in regulatory approval and the commencement of operations and operations blockaded by communities. Conversely, good relationships secure efficient operations and financially material savings.
- Enhance reputation and therefore 'social licence to operate'.
- Increase 'regulatory goodwill', which could lead to faster permitting.
- Influence emerging environmental regulation and policy.
- Reduce costs of compliance with environmental regulations (because it may be more cost-effective to offset than to mitigate, and because best practice on biodiversity can speed permit negotiations etc., as above).
- 'First mover' advantage for innovative companies: as biodiversity offsets take off, leading companies will be in a position to shape the development of new markets for offsets or expertise in biodiversity offset design and implementation and secure a niche in emerging markets.
- Maximize strategic economic opportunities in emerging markets (e.g. establishing companies to implement offsets).

advantage as preferred partners with governments, funding institutions and other stakeholders. Conversely, bad environmental practice can lead to higher operating costs, costly permit delays, liabilities and lost revenues (Grigg and ten Kate, 2004; Foxall et al, 2006). Box 13.2 lists a full suite of motivations for undertaking voluntary biodiversity offsets.

The 'business case' for multinational companies undertaking voluntary biodiversity offsets globally is often quite different from the estimations of return on investment of those involved in specific wetland and conservation banking business opportunities in the US. The policy framework in the US has created a regulated market that has enabled mitigation and conservation bankers to estimate the return on specific investments in listed wetlands or habitat for listed species. By contrast, a multinational mining company considering a voluntary biodiversity offset for a new mine in a developing country will estimate the financial commitment involved in designing and implementing the offset and weigh this against the likely business benefits, but will rarely be in a position to calculate specific return on investment. Implementing the offset will probably be one among an array of strategies that helps the company secure licence to operate, to operate efficiently on good terms with local communities and regulators, and to avoid costs and liabilities associated with poor practice.

Barriers to offset uptake

A 2004 report published by Insight Investment and World Conservation Union (IUCN) interviewed 50 individuals from companies, governments and conservation organizations on the business case for biodiversity offsets (ten Kate et al, 2004). The report concluded that while there is widespread interest in the mechanism of biodiversity offsets, certain barriers have prevented the mechanism from being used more widely. There is little practical experience with the mechanism of biodiversity offsets beyond the regulatory regimes described in Table 13.1. Business, government and the conservation community need to see how biodiversity offsets work in a wide range of circumstances to assess their effectiveness in various industry sectors, and for a range of biodiversity impacts, for operations at different scales, in different regions and ecosystems and policy environments.

Also, there is currently no single source of accepted methodologies and guidelines on how to design and implement biodiversity offset projects. This is something the Business and Biodiversity Offset Program (BBOP) introduced below has been established to address. Companies are keen to ensure that their voluntary efforts are regarded as socially acceptable and scientifically credible. For this reason, private sector representatives have asked for broadly accepted guidelines that will lend credibility, practicality and political support for the approach.

Challenges for offset design

In addition to the lack of experience and accepted guidelines, there are significant challenges inherent in designing biodiversity offsets.

First, there is concern that offsets will pave the way for authorities to grant planning permission to projects that would otherwise be turned down because they are too damaging to the environment. Offsets must be designed within the context of the mitigation hierarchy, once the appropriateness of a development project has been determined. Some projects would bring about such severe biodiversity and environmental impacts as to be inappropriate. Even where it has been granted government approval, developers planning projects should carefully consider the potential impact on its reputation and licence to operate before proceeding. Some projects may be granted government approval for reasons of overriding public interest, even though they will make a very serious impact on biodiversity. There is growing consensus that in cases where an impact simply cannot be offset (for example, where the project makes a locally endemic species extinct), any compensatory conservation activities could not properly be termed an 'offset', although they may well be encouraged, since the development project is inevitable.

Once the initial hurdle of determining whether a project and a biodiversity offset are suitable has been overcome, issues surrounding offset quantification and design arise. Perhaps the most fundamental challenge in designing offsets is how to quantify 'no net loss of biodiversity' and find a currency that will measure impact and determine that an offset is suitable and adequate. Central to achieving no net loss is the ability to demonstrate that whatever land management changes have taken place – whether they involve restoring a degraded site or averting the risk of loss – generate conservation outcomes that are additional to what would have happened without the offset. It is also important that land management changes on one piece of land do not displace unsustainable activities to another piece of land.

Another challenge is to establish and measure the limits of a project's impact, so the appropriate scale of the offset can be determined. While it may be comparatively straightforward to quantify a project's direct impacts on biodiversity (i.e. caused by the construction of a mine pit, roads, infrastructure, etc.) there is significant uncertainty about how to measure and offset indirect impacts (e.g. the impact on biodiversity brought about by the inward migration of people that is often associated with a new mine). Indirect impacts are generally not the sole responsibility of the project developer and it may be difficult to determine the boundary of responsibility for them. Solutions may lie outside the developer's sole control, necessitating partnerships between developers, governments and communities.

Lack of consensus on priorities for biodiversity conservation can also be difficult for offset planners. Indeed, they may be deterred from undertaking a voluntary biodiversity offset by the risk of criticism of the offset activities they select.

Local communities, whose use of natural resources will often be the determining factor in the successful implementation of a biodiversity offset, need to be motivated to support the offset over the long term. Central to the success of offsets is thus that there is an equitable distribution of the economic, social and environmental costs and benefits of offsets and that the concerns of local communities are integrated into offset design from the outset. Development projects often have a negative impact on local people's access to biodiversity (for instance, people may no longer be able to access the project area for hunting, fishing, gathering forest products or recreation). Offsets should aim to ensure that local communities' use of biodiversity is restored or compensated. In addition, the success of the conservation activities on the offset site(s) is likely to be affected by the underlying causes of loss of biodiversity at those sites, which may in turn be a result of activity by local community members. Local people's needs for food and fuel wood for their livelihoods, for instance, may be a reason that biodiversity is being lost at the offset site. For the offset to succeed, therefore, it will be necessary to work with local communities to help them meet their biodiversity needs in more sustainable ways. Even more fundamentally, any offset site and activities chosen must not have a negative impact on community biodiversity use in themselves. If sites are chosen in collaboration with communities, such a scenario should not arise, which again underlines the importance of a participatory approach to biodiversity offset design.

In 2005 an international initiative was created to address these many challenges and to develop internationally accepted standards of best practice on biodiversity offsets that enjoy support from companies, governments, communities and environmental groups.

The business and biodiversity offset program

BBOP is a new international partnership of companies, government agencies, scientists and non-governmental organizations (NGOs). The BBOP partners wish to show, through a portfolio of pilot projects in a range of industry sectors, that biodiversity offsets can help achieve significantly more, better and more cost-effective conservation outcomes than normally occur in the context of infrastructure development. The BBOP partners also believe that demonstrating no net loss of biodiversity can help companies secure their licence to operate and manage their costs and liabilities.

The objectives of BBOP are:

- To achieve conservation and livelihood outcomes in a portfolio of diverse biodiversity offset pilot projects worldwide. The pilots aim to demonstrate 'no net loss' (or preferably 'net gain') of biodiversity and benefits to local communities.
- To compile, use and disseminate a toolkit for designing and implementing best practice biodiversity offsets. The programme will draw on the experiences of the pilot projects, the advisory committee and learning network to develop the toolkit.
- To influence developments in legislation, policy and corporate practice related to biodiversity offsets so that they meet both conservation and business objectives.

BBOP[1] is working with companies in several industry sectors, and policy makers globally, to ensure that all new operations integrate a lasting conservation component to offset the impact on biodiversity. The offset conservation activities will protect threatened habitat and contribute to national biodiversity strategies. The involvement of local communities in the design of biodiversity offsets is particularly important. One of the principal business motivations of the developers in undertaking the pilot projects is to ensure that their local stakeholders feel that the project has delivered not only economic benefits such as jobs, but also has not adversely affected their living environment, and, preferably, improved it. For this reason, the pilot activities will involve and aim to benefit local communities and, to the extent possible, work with experts from local NGOs and universities.

The BBOP partners are selecting pilots to cover many factors, including marine and terrestrial ecosystems, geographical diversity, scales of operation, industry sectors and policy settings. This varied portfolio will allow BBOP partners to assess how these factors influence the success of biodiversity offset projects.

BBOP has established an International Advisory Committee, comprising experts from companies in different sectors and from government departments, taxonomic, conservation, research and academic organizations worldwide. These experts are drawn from disciplines that underpin biodiversity offsets, including: conservation methodologies and metrics; bioregional and landscape scale planning; systematics and biodiversity measurement and monitoring; risk, project and biodiversity management in business; and environmental economics. Many have already assisted in designing and implementing biodiversity offset projects and the associated public policy. This group is preparing a toolkit of best practice on biodiversity offsets that will include and influence emerging policies on biodiversity worldwide.

In parallel, BBOP has established a learning network of companies, industry associations and government representatives and financial institutions to participate in learning, exchanging experience and promoting best practice in biodiversity offsets. The learning network provides the forum for broad debate on biodiversity offsets.

The vision and expectation of the programme is that biodiversity offsets will become a standard part of business practice for all projects with a significant impact on biodiversity. The routine mainstreaming of biodiversity offsets into development practice should result in long-term and globally significant conservation outcomes.

Future opportunities and next steps for biodiversity offsets

The growth of corporate practice, regulations and loan conditions requiring biodiversity offsets suggest that more developers will use them in the future to manage their risks and create business opportunities. A number of governments are planning to introduce or revise policy on biodiversity offsets or 'compensation', so that developers are required to take full responsibility for their footprints in a more quantified manner than in the past. International best practice on biodiversity offsets will emerge in the next five years from experimental programmes involving all the stakeholders, such as BBOP.

Intergovernmental policy on biodiversity offsets is also on the cards. In March 2006, the United Nations Convention on Biological Diversity (CBD) adopted a decision on engagement of the private sector, which includes text on biodiversity offsets:

> *Noting that contributions from business and industry towards the implementation of the Convention and its 2010 target could be facilitated by further work under the*

> *Convention to develop: … (c) Guidance for potential biodiversity offsets in line with the objectives of the Convention. (CBD, 2006)*

This decision is a signal to national governments that biodiversity offsets are a mechanism to be explored as a practical means to achieve parties' commitments to significantly reduce the rate of loss of biodiversity by 2010. This could take the form of guidelines for companies, governments and others on biodiversity offsets.

We envisage a number of developments in the coming years that, together, will mean that biodiversity offsets are routinely factored into projects and increasingly become an expectation of project developers:

- A set of principles to guide the development of voluntary biodiversity offsets.*
- A set of 'Key Questions', widely accepted that any developer purporting to undertake a biodiversity offset should be able to answer plausibly and back up with evidence.*
- A set of accepted and cost-effective methodologies for quantifying projects' impacts on biodiversity and designing suitable offsets.*
- A clear business case, based on the advantages to companies of undertaking biodiversity offsets, documented through practical experience.*
- A range of different pro-offset policy models in use by governments, ranging from highly regulated market-based systems such as conservation banking in the US, to more creative use of existing regulations, such as the regular use of EIA and planning processes to require offsets.
- Local, and possibly regional, markets for biodiversity that support rational land-use and systematic conservation planning.

> Items marked with an asterisk are developments the Business and Biodiversity Offset Program is working on in its first phase.

Biodiversity offsets are likely to become more sophisticated than compensation projects have been to date. For instance, future offsets may well be 'composites' of different activities, generating benefits for different groups involved. One component of such a composite offset is likely to be local to the development project's impact. This component would typically aim to maintain adequate ecosystem services in the project impact area and ensure local communities' use and enjoyment of biodiversity was not adversely affected. Another component of the offset could be further afield. It could support better land-use planning at a landscape and even regional scale and contribute to national and global conservation priorities. This aspect of the offset could be designed to secure a representative sample of all biodiversity patterns while maintaining the ecosystem services upon which the world's growing population relies. It could create conservation corridors and larger conservation areas that will be more resilient in the face of global changes, also generating economies of scale and paving the way for some kind of conservation banking.

Altogether, biodiversity offsets have great potential to help both developers and conservation in the coming years. But first things first: our emphasis in the coming years is to produce replicable, cost-effective methodologies, experiment with pilot projects and communicate the results; and collaborate with policy makers to introduce policy frameworks conducive to business and to biodiversity.

Note

1 BBOP is managed by Forest Trends and Conservation International. www.forest-trends.org/biodiversityoffsetprogram

References

CBD (2006) Report of the eighth meeting of the parties to the Convention on Biological Diversity, United Nations Environment Programme, Decision VIII/17, p259 www.biodiv.org/doc/meetings/cop/cop-08/official/cop-08-31-en.pdf

F&C Asset Management reports on biodiversity (2005) See www.fandc.com/newsDetail.asp?newsID=348 and www.fandc.com/new/aboutus/Default.aspx?id=63880

Foxall, J., Grigg, A. and ten Kate, K. (2006) *Protecting Shareholder and Natural Value: 2005 Benchmark of Biodiversity Management Practices in the Extractive Industry*, Fauna and Flora International and Insight Investment Management Limited, available at www.insightinvestment.com/Documents/responsibility/Reports/protecting_shareholder_and_natural_value_2005.pdf

Goldman Sachs (2004) 'Global energy: Introducing the Goldman Sachs energy environmental and social index', *Energy Environmental and Social Report*, 24 February

Goldman Sachs (2005) 'Global energy: Sustainable investing in the energy sector', www2.goldmansachs.com/

Grigg, A. and ten Kate, K. (2004) *Protecting Shareholder and Natural Value. Biodiversity Risk Management: Towards Best Practice for Extractive and Utility Companies*, Insight Investment Management Limited, London

McKenney, B. (2005) *Environmental Offset Policies, Principles, and Methods: A Review of Selected Legislative Frameworks*, Biodiversity Neutral Initiative, Nanaimo, British Columbia

Miranda, M. et al (2003) *Mining and Critical Ecosystems: Mapping the Risks*, World Resources Institute

ten Kate, K. (2003) 'Biodiversity: Towards best practice for extractive and utility companies', Insight's presentation to the World Parks Congress, 13 September, http://www.insightinvestment.com/responsibility/project/biodiversity.asp

ten Kate, K., Bishop, J. and Bayon, R. (2004) *Biodiversity Offsets: Views, Experience, and the Business Case*, Insight Investment and IUCN, London

14

Australia's Biodiversity Credits

James Shields

Introduction

Australia is an egalitarian country. This sense of equality extends to the land, its plants, animals, rivers, oceans and mountain ranges (although the beach might get an extra tick among landscape features). Australia also has one of the oldest national park systems in the world and many would testify that wild spaces are part of the country's national identity. Unfortunately, European and Aboriginal settlers did not have complete knowledge and hence opportunity to treat all creatures, plants and landscapes in an equitable manner (Clark, 1969). Consequently, many of Australia's landscapes have not been protected or preserved and a variety of ecological communities are now listed as threatened. Fortunately, economic forces are now positioned to support the landscapes that make Australia unique.

Biodiversity banking in the state of New South Wales (NSW) is now a reality, and proponents hope it will soon allow ecologically sustainable development to reap an economic reward. In NSW, a bill establishing the legal framework for biodiversity banks was introduced into parliament in May 2006. The regulations in the Act allow developers who need biodiversity credits to acquire them, and biodiversity bankers and brokers to sell them. Although some of the first definitions of biodiversity assets, debits, credits and units came from Australia (Shields, 1997; Gibbons et al, 2002; Brand, 2005), banking and trading in these entities has taken place only on an informal basis through offsets or payment for services (feral animal control, fencing streams, destocking, replanting). The banking scheme lags behind the US at present but, if successful, stands to create a template and standard that can be used extensively around the world to provide a means of 'making the priceless valuable'.

Historical context

Australia is an isolated island continent with a unique flora and fauna. In the absence of humans and their associated pests (e.g. domestic animals, rats and mice), Australia played host to the evolution of animals such as marsupials (pouched mammals), monotremes (egg-laying mammals), few placental mammals (rodents, bats, dingoes) and a disproportionately high diversity of hollow-nesting birds. Australia has a wide variety of terrestrial and probably even more aquatic ecosystems, the latter due to the high ratio of coastline to land area. This chapter will focus on terrestrial ecosystems and associated freshwater habitats.

The arrival of Aboriginal people sometime between 50,000 and 20,000 years ago coincided with a number of changes in Australian biodiversity: the mega-fauna (giant wombats and kangaroos), carnivores down to dingo size, and rainforest communities disappeared or decreased dramatically, and were replaced by eucalypt-dominated landscapes, vertebrate communities featuring medium to small size species, and very small carnivores. The larger thylacine and Tasmanian devil survived on Tasmania, where there were no dingoes – as did major rainforests; the Aboriginals there had lost the ability to make fire, although they used it when it was available. As well as increasing mobility, fire had the added advantage of creating new grass growth which attracts large herbivores (kangaroos and wallabies), a major source of food for Aboriginal populations. By moving around to appropriate parts of the landscape, Aboriginals had established a relatively stable population and way of life when Europeans began to arrive about 230 years ago.

Exploration and settlement by Europeans occurred relatively late in human history, although the northern parts of the continent were regularly visited by people from south-east Asia for several hundred years. The unstable, dry climate and low fertility of the soil has prevented intensive development in all but the coastal fringes. Consequently, although Australia has a large land mass (approximately 10 per cent larger than the continental US), the human population remains small (less than 20 million people). It could therefore be expected that natural resources – including biodiversity – should be well preserved and relatively pristine. Such is not the case. Although *intensive* development was confined to the coastal fringe, large proportions of the continent have been severely degraded by *extensive* land-use practices (grazing sheep and cattle, water use for dry-land farming of rice and cotton). Historically, these agricultural and pastoral practices were developed for areas with fertile soils and plentiful rainfall (northern Europe and parts of Asia). In combination with the introduction of biological pests (in particular rabbits, foxes, cats and rodents), extensive land use by humans has degraded the quality and quantity of natural resources across inland Australia. Until recently, the dry inland areas where these practices occurred were not considered 'threatened' because they were devoted to established agricultural practices regarded as essential to the national economy. Consequently, conservation initiatives and action were not focused on the vast majority of the continent and its most threatened biodiversity.

Instead, over the past 50 years, conservation action has concentrated on the forested areas near the coast. Many inappropriate forestry practices, particularly clearing native eucalypt forest to make way for northern hemisphere pines, were utilized early in the 20th century, and attracted justified criticism from environmental agencies and groups. The focus for acquiring land for the national park system remained clearly on the forested areas of the continent, and to a certain extent still does in 2007. Early conservation initiatives established biosphere sized reserves in the tropical rainforests in the north-east and the alpine areas of the south-east prior to 1990. Conservation focus on forests intensified around the country at this time, and a multi-government policy was initiated to resolve conflicts, establish a national system of forest reserves and provide for a sustainable timber harvest. The actions to achieve these goals were set out in the National Forest Policy (NFP) of 1992. These actions provided the biodiversity data, the knowledge of land-tenure distribution, the technical capabilities and the socio-economic information necessary to establish biodiversity markets in Australia.

Regional forest agreements and negotiated partnerships: Providing tools to build biodiversity banks and markets

The process of implementing Australia's NFP was completed by 2002. Large areas of production forest were converted to the reserve estate following Comprehensive Regional Assessments of biodiversity and timber resources. These assessments provided information for an objective negotiation process resulting in Regional Forest Agreements (RFA). The purpose of RFA was to provide a binding and equitable document which would resolve the conflict between nature conservation requirements and timber harvest. RFAs last for 20 years after agreement, with a review every five years.

Two key elements that were necessary in RFAs are also necessary for implementing biodiversity markets. The first was achieving a resolution regarding the policy and criteria for managing the forest estate agreed by the state and Australian governments. This gave the necessary direction, authority and impetus to resolve biodiversity management issues successfully. The policy clearly set out that 'a Comprehensive, Adequate and Representative Reserve (CARR) system' would result from RFAs for each state. An expert working group set out the criteria for such a system, which included a definition of the word 'adequate' in terms of area reserved. The minimum reservation area for adequacy was set at 15 per cent for each forest ecosystem (Shields, 2003). This effectively set targets, which could then be used for negotiation in terms of land area set aside for forest reserves versus timber resource lost. These targets provided a means of measuring success, setting priorities for action and determining relative economic values for different elements of biodiversity. This sound background in policy, regulations and economics provided the necessary environment for the development of biodiversity markets on their own.

The second important element with regard to biodiversity markets was a means of measuring the asset – that is, the biodiversity – and thus progress towards the targets for biodiversity conservation. In NSW, the negotiations for RFAs used hectares of forest ecosystems as the surrogate for biodiversity (e.g. standard forest types or vegetation associations) in measuring progress towards the goal of 15 per cent within the CARR system. The NFP also requires that needs of threatened species be given direct and primary consideration in establishing a CARR. This was done in NSW by using the area required for a minimum viable population of each threatened species relevant to the region. Putting hectares of threatened species or communities into the reserve system was regarded as progress towards the target of 15 per cent representation. In addition, forest condition (old growth versus regrowth) and wilderness values were used as measures in reserve selection (as required by the NFP). To analyse all of these values, a conservation planning tool was developed by Bob Pressey and Simon Ferrier of the NSW National Parks and Wildlife Service, called C-Plan. This computer package allows land units (200ha compartments were used in the NSW RFAs) to be evaluated according to the biological, cultural and physical attributes that occur in that unit. These units are then grouped to give the maximum conservation value. The results of different scenarios can be compared to assess different environmental, economic and social outcomes. C-Plan can be used to provide a direct assessment of biodiversity values (Pressey et al, 2000) across extensive landscapes.

Under the present regulations for biodiversity banking in NSW, the credit/debit is measured by collecting environmental and abiotic data for the area under consideration and then applying a rule set similar to that in C-Plan. In real terms, the land under consideration for development is evaluated for the presence of threatened species habitat, threatened species, threatened ecological communities, development impacts, and in terms of connectivity with the local and regional reserve system. A value is then calculated in terms of the quantity (hectares) and type (usually a specified vegetation community) of biodiversity credit that must be acquired for the development to be approved. For example, a development for senior citizen housing on the south coast of NSW has been provisionally approved that will require clearing about 20ha of forest with a 60ha block, which currently provides habitat for the powerful owl (*Ninnox strenua*). There is significant retained vegetation within the development (40ha), and connecting corridors with local reserves are part of the development plan. Due to the large amount of retained habitat and the connecting corridors, the offset figure has been set at three to one, a relatively low figure. In real terms it means that the developer must acquire 60ha of habitat suitable for the powerful owl (*Ninnox strenua*) occurring in the same geographic region and vegetation community. A less sensitive development plan might have attracted an offset ratio of ten to one to achieve a similar conservation outcome. If the development proposal were to clear the entire 60ha property and the offset ratio was then set at ten to one, then a conservation offset of 600ha would be required.

Resolving further the issues confronting Australian biodiversity

RFAs were negotiated for all of the coastal forests in NSW, the Alpine area west of the Snowy Mountains, and the Brigalow Belt South in central NSW between 1997 and 2002. Using C-Plan, a best-possible CARR was put in place for these regions. Unfortunately, the process could only consider forest ecosystems on publicly owned land. Biodiversity, particularly threatened biodiversity, requires management across the entire landscape, and whole ecosystems were missed out during the forest negotiation process. Consequently, most forest communities are relatively well reserved and most of the individual species that make up these communities are relatively well protected. Although the issue of adequacy remains debatable, the NFP and consequent RFAs undoubtedly established a sound basis for conservation of forest biodiversity, the issue which had dominated conservation action in the past.

Biodiversity outside the forest ecosystems has not been through the same process, and the land is largely in private ownership. Private lands are generally managed to create wealth for their owners. If the biodiversity has no value, the most practical course of action is to remove it and create something that makes money.

This leads directly to a need for market forces that create wealth from positive management of biodiversity. However, this concept did not exist in 1992. Resolving the forest issues was the most important conservation item on the Australian agenda up until the turn of century. There was no particular pathway to further resolution.

Finding a way forward: Conceiving and defining the value of a new natural resource

Finding a new pathway to solve the dilemma facing biodiversity management outside the command control system of making national parks out of state production land in NSW was enabled by serendipity. Among other things, a Canadian named David Brand had been convinced to take an executive job with State Forests of NSW – as it was called then – to negotiate and resolve the contentious environmental issues that faced the agency, particularly resolving the RPAs. Dr Brand had just come from successful negotiations involving partnerships and sustainable forest management in Canada. His worldwide contacts, positive viewpoint and scientific background were a much needed inspiration. The Eden forest management team convinced him to come down on a field trip shortly after arrival in NSW. One morning he came in to the office and checked his email (a relatively new convention in Eden in 1996), and said, 'Hey, look at this – the Costa Ricans are selling air!'

This was the beginning of the carbon credit market in Australia and NSW. David Brand and others, notably Bob Smith and Kim Yeadon, developed a carbon trading scheme in NSW that was the first of its kind in the world, and carbon credits became a reality through Kyoto in the next decade.

If you can sell air, it occurred to me at the time, you can surely sell biodiversity for a reasonable return. David Brand, Bob Smith and others encouraged thoughts along these lines – particularly the thought of accounting for biodiversity in some reasonable manner.

With that concept – economic value for biodiversity – I joined those elsewhere who were attempting to put a real value on biodiversity. In Australia, Drs Sue Briggs and Phil Gibbons were collaborators in conceiving positive solutions for natural resource management.

From forests to the entire landscape

Following the RFA process (2002), there was little public land available from which to create new reserves and, as previously mentioned, forest biodiversity was relatively well protected. Threatened species were concentrated in the sheep–wheat belt, arid regions, wetlands, riparian areas, estuaries and highly developed coastal forests. Key threats concentrated around continued clearing in these habitats due to expanding urban areas, agricultural intensification (clearing native vegetation), feral animals and salinity/climate change.

The (non-forest) system for land development in NSW requires environmental assessment, impact statements for development, and approval by state and local government. There is an overarching national regulation, the Environmental Protection and Biodiversity Conservation Act, which is only relevant on land owned by the Australian government or in special cases relevant to national or international conservation requirements (unlike the US regulations, which apply to most lands). Applications for clearing in NSW are approved with conditions to ameliorate detrimental impacts on biodiversity. These conditions usually involve either retaining some of the biological values on site, or some negotiated offset. This system has been in place for the past 20 years in NSW, has had little positive effect with regard to managing biodiversity and has become increasingly cumbersome to implement. Habitat for threatened species has been preserved in a random manner, at best, and costs to developers are high. It was at this point the question became, 'How could market forces combine with ecological management (biodiversity banking) to deal with ecological degradation in NSW?'

For instance, the squirrel glider (*Petaurus norfolkensis*) is listed as 'vulnerable' in the Threatened Species Conservation Act (TSCA). Development in coastal woodlands and forests, particularly in the Hunter Valley, north of Sydney, threatens local populations. Despite current regulations and the conditions imposed on development approvals, individual animals and family groups continue to be displaced by housing and other developments in the area. They are one species that make up the forest and woodland ecosystems that are in decline due to continued human pressures. It is possible – using ecological management techniques and careful planning – to maintain these communities in the context of human development, but the current regulations in NSW are not achieving success with regard to individual species or ecological communities. It was simply too complex and expensive to use command control regulations to manage the squirrel glider successfully.

Ecological economics, threatened species banks and other major initiatives to change the way we manage natural resources were happening simultaneously (1998–2002). It became apparent that to deal with biodiversity in economic terms, property rights would have to be defined, markets analysed, buyers and sellers identified, and ecological entities (communities in particular) defined. Further, contractual and fiduciary arrangements would need to be clearly spelled out in legal and enforceable terms. Finally, to drive the market, there would need to be some regulation or legislation that compelled trade.

An equation to define bios (a unit of biodiversity) grew out of a collaboration between ForestsNSW and Macquarie University, sparked at an agricultural economics conference (Shields, 1997). This was made possible through the transaction called Tuan Trade-off, in which the bios equation was used to offset a loss of 10ha of tuan (*Phascogale tapotafata,* a small insectivorous marsupial) habitat in the present by securing 100ha of habitat for the future. Field surveys, community liaison and real estate bargaining was carried out to derive the raw figures and areas of habitat required and available. Unfortunately, due to objections from local NGO action groups, the deal was not completed. A system was emerging which could work successfully if it were transparent and acceptable.

One of the main objections to biodiversity banking arrangements is that there is no regulatory or operational system that will guarantee that the banked areas are managed successfully. The RFA process provided further essential work with regard to this issue. As part of each NSW RFA, a contractual arrangement for managing biodiversity was negotiated and field tested, and these licences have been implemented over the past five years (the Threatened Species Licence in all NSW RFAs). These licences are regulated by the NSW Department of Environment and Climate Change (DECC), and NGOs as well as the general public can and do participate in this regulation (e.g. the Green Police who voluntarily inspect logging areas).

Other economic, social and technical developments pertinent to biodiversity banking were addressed. Various offsets and investment schemes were developed, and some successfully implemented in NSW. In Victoria, work to define a biodiversity unit (the habitat hectare) was implemented to allow efficient government investment in biodiversity management. This was completed successfully in the Bush Tender, a reverse or 'dutch' auction conducted by the Victorian government. Victorian landholders were invited to tender for funds to improve their biodiversity assets – for example, bid for grants to plant trees, fence remnant vegetation or destock their land. This proved quite successful, transparent and effective.

In Western Australia, a partnership between World Wildlife Fund and the state government has produced a marketing scheme for biodiversity based on increased real estate value, conservation interest, tax incentives and opportunities to exploit sustainable resources (honey, wild flowers, ecotourism, among others).

Biodiversity banking is currently most well developed in NSW, where a team from DECC (2005) has developed computer tools (grown out of C-Plan) to allow biodiversity assessment and definition in the context of clearing native vegetation. In NSW, the Native Vegetation Act (NVA) and the Threatened Species Conservation Act (TSCA) were passed, which explicitly prevent unlicensed clearing of native vegetation or harm to the biological entities listed on schedules of the TSCA. These tools and regulations are key in enabling a biodiversity bank for technical and economic reasons (e.g. there is to be no net loss of native vegetation, and biodiversity can be accounted for in a transparent manner.

In combination, these factors made possible the development and implementation of a biodiversity management scheme (Biodiversity Banking) that will create economic value from its operation. That is to say, it will be possible to buy and sell biodiversity if anyone wanted to do so. The means to define the property rights of biodiversity are through listed and defined Endangered Ecological Communities (EECs), habitat areas for minimum viable populations and conservation planning tools. There are tested contractual arrangements for biodiversity management (Threatened Species Licences), financial arrangement and contracts to ensure that resources are available to meet management requirements and provide for a return to biodiversity bankers described in the regulations for the Biodiversity Banking Act. Most importantly, there is regulation in place which compels the market; for example, no clearing of native vegetation (NVA), and no net loss to listed biodiversity (TSCA).

Issues and answers for sustainable biodiversity management

Dealing with success: What is it?

The status quo of Australian biodiversity at the turn of the 20th century meant that it was now necessary to successfully manage biodiversity across the agricultural, industrial and urban parts of the landscape. As this is where the majority of endangered taxa occur, where the most ecosystems have been converted to anthropogenic uses, where landscapes are dysfunctional, poisoned or threatened with both, and where ecology meets the economy very directly, previous solutions will not work. First, there are no large portions of government-owned land available to convert into reserves (Parks) or manage in a sustainable, multiple use fashion (State Forests). Second, land uses in the private sector are not as amenable to sustainable multiple-use practices as is forestry. Building houses, factories, transport systems and growing crops requires conversion of land from native vegetation. The current practice of requiring *in situ* preservation (for instance, keeping or rehabilitating 15 per cent of the native vegetation on site) is not effective ecologically and is very costly in financial terms. This is undoubtedly so because there is no coordinated plan of management for these areas, there is no transparent and effective process for selecting them and there is no assurance that *in situ* reservation or ad hoc offsets will produce any outcome with regard to successful biodiversity management.

Between 2003 and 2004, the Department of Environment and Conservation in NSW and other government agencies set out on a programme to achieve the state's policy to maintain or enhance all biodiversity and abate identified threats. Key Threatening Processes (KTP) had been set out and listed on the schedules of the TSCA. Once a KTP is legally registered on the schedules of the TSCA, the preparation and implementation of a Threat Abatement Plan (TAP) is required. A new regional administrative unit was created, the Catchment Management Authority (CMA), to process regulations and plan government actions with regard to natural resource management, including biodiversity, at this time. Based on geophysical boundaries, these units should be relevant base areas for managing biodiversity and water.

Given the above, it is possible to set out a long-term conservation plan, which defines the amount of land area, prioritizes biodiversity values and prescribes management conditions that will give a successful, sustainable and long-term result (the equivalent of a forest CARR system). With such a plan in hand, the CMAs can use existing rules to deliver the result, in conjunction with other government agencies and biodiversity assets (Park and State Forests).

Such plans – regional conservation plans, local environmental, catchment management 'blue prints' – have been or are being prepared across the state at present (DECC, 2006). For instance, the conservation plan for the Lower Hunter region just north of Sydney is to be released in October 2007.

A vision for success in biodiversity management can be summarized thus:

- Ecological communities that comprise regional landscapes are managed through long-term binding agreements and occur across at least 15 per cent of their pre-1750 distribution. In semi-arid regions, it is reasonable to increase this to 30 per cent.

- The management agreements contain an integrated monitoring programme that is well funded, and considers biodiversity across the entire range of biota in accordance with its importance in ecosystem function.
- Threatened species and populations are identified and managed through effective and fully financed programmes that maximize recovery opportunities.
- Key threatening processes are identified and described, as are the most efficient management actions for the abatement of these processes.
- All management actions are monitored sufficiently to guide adaptations to the planned course of actions based on results.

The money

The problem that remains, as always, is the money. To carry out the activities above and to acquire new biodiversity assets (land) requires funding. Although this funding is small compared with spending on military, infrastructure and health programmes, it is large when compared to the current budget for natural resource management. A simple way to redress this would be to change the priority of government spending. When asked how to solve the ecological 'crisis' of the late 1980s, Jack Ward Thomas (then the Chief of the US Department of Agriculture Forest Service) said: 'It's easy – build one less B2 bomber.' (Thomas, J. W., pers. comm., 1989). Although simple, this course of action does not seem likely. Military spending and other, similar issues remain top priority for most government spending.

Another way to fund biodiversity management is to develop a successful market that allows owners to sell and buyers to purchase valuable biodiversity assets. As with all markets, this would require market drivers – that is, economic, social or political forces that induce the exchange of funds. At present there are no forceful economic or social factors driving a market in NSW (although economic drivers could exist – it makes as much sense to buy and sell biodiversity as it does to trade in rare stamps, fine art or Barbie© memorabilia). Therefore, the conclusion has been that biodiversity markets will only be truly effective if there are political forces (regulations) that require developers and land users to transparently demonstrate that they maintain or enhance biodiversity values (e.g. their ecological accounts balance or are in profit). In NSW, all of these elements including regulations to compel actions were in place by 2003. However, there was no coherent plan or process by which these compelled actions could be organized or funded to effectively produce biodiversity outcomes.

It then required the bold step of developing policy or regulation or both to establish a biodiversity bank itself. This exactly what Simon Smith and his team have done over the past three years. The bill was passed by the NSW parliament (2006) and it establishes biodiversity banking rules. Regulations have been drafted and implementation procedures started in 2007. The Act and its regulations set out processes for establishing a biodiversity bank, for assessing the biodiversity values in banked areas, for assessing biodiversity assets in development areas that require maintenance or enhancement, and a transparent scheme whereby developers may purchase the appropriate type and amount of biodiversity credit. Biodiversity brokers and biodiversity assessment criteria are established by the legislation. Regulations for implementation are in development by four focus groups. If the market succeeds, the money for finalizing success in biodiversity management should be available in sufficient quantity to meet the targets.

Markets and marketing

If there is to be a market solution for biodiversity management, the basic factors required and the problems that are probable should be considered. We do so in the following sections, with the intention of defining and resolving problems that could arise.

Market forces and economic drivers

It is relevant to explore the general question: What are the market forces that will drive the creation of wealth by maintaining or enhancing biodiversity?

Over the decade from 1996 to 2006, we have entered into many formal and informal debates on the value, nature and implementation of a scheme of economic rewards for biodiversity. Without fail, the debate comes to a discussion of market forces, and the more specific question: Who would participate in such a market?

We can describe six types of participants:

1 Those who are required to by regulatory or economic forces. In the Australian example above, anyone who wants to clear native vegetation or harm a listed biological entity has to show that their activities will maintain or enhance that resource.
2 Those who will derive some value from it – for example, investors who think they may get a reasonable yearly rate of return on their investment.
3 Those who will derive some advertisement or public good out of it – improving corporate image and tax benefits.
4 Individuals who want to own the assets for its intrinsic value and the property rights created in a biodiversity in the bank. Collectors, interested groups and other entities who value the asset as a possession.
5 Philanthropist or NGO investments to achieve primary goals.
6 Those who can combine biodiversity banking with other enterprises (eco-tourism, wildlife use, recreational businesses such as mountain bikes or abseiling).

Will the Biodiversity Banking Bill in NSW successfully engage the six types of customers listed above? The simple answer should be 'yes', because it should be economically advantageous to participate in biodiversity banking rather than any other regulatory process for development approval (Participant 1). Other types of investors and biodiversity 'developers' (Participants 2–6) may engage also, but there are less compelling forces for participation. Their participation will depend on successful public relations and marketing, efficient processing by agencies, entrepreneurs (Participant 6), and at least some positive examples of successful conservation through development banks.

Market operation

If sufficient economic forces are present to create a market, there still remain considerable problems with initiating and operating such a market. Indeed, markets may not be the answer in NSW. The issue is to acquire the money to fund a course of action for resolving biodiversity issues. There are many tools besides a hammer, and many outputs other than a driven nail. The cost of transaction may prohibit a market developing; there may be no effective institutions to provide a secure basis for the exchange of funds, the relative values of the biodiversity are difficult to establish, and there can be problems with transparency, security and delivery.

However, none of the factors listed above prohibits the operation of a market. Humans have been buying and selling plants, animals and groups thereof (biodiversity), vegetated land (habitat) and management services (grazing leases, mustering contracts, timber concessions) for thousands of years. Biodiversity assets have the same basic characteristics as livestock, fields and forests. If they are truly valuable, biodiversity assets and management contracts can be handled just as simply as any other tradable natural resource.

Although it is clearly possible to trade in biodiversity, the details are new and different, requiring definitions, principles and processes to be created. The detail required may seem overwhelming; indeed, many theoretical economists and ecologists agree that it is so complicated that we might as well not begin. Similarly complex problems occur in society, and are dealt with through the legal system and lawyers. Shakespeare noted, to form the perfect society we must first kill all the lawyers. We would note here that to start a biodiversity market, we must first kill (metaphorically) all the economists, and then all the ecologists. My point (and Shakespeare's) is that if we ask the experts who are responsible for the current system to repair it, these experts will most probably repeat past mistakes, argue to defend the past system and fail to start innovative measures to resolve the situation. It is necessary to seek and accept new solutions for resolving conservation issues.

Dealing with success: How could we fail?

It is always useful to consider obstacles to a proposed course of action. Over the past decade, it has become apparent that there are specific and recurrent contentious issues. Through research, trial and error, and 'enlivened' discussion, it has been possible to define recurrent problems, obstacles and difficulties for biodiversity banking. These are set out as 'issues' below. Through the same processes (research, trial and error, enlivened discussion), it has been possible to derive and set out some 'resolutions'.

Issue 1: Failure to set or meet targets for success, and the associated risk of failure through lack of an exit strategy are recurrent issues in biodiversity banking. Unrealistic or unachievable biodiversity targets can result in a very large area requirement or biodiversity debit to cancel a very small biodiversity credit. Another description of this issue is that the high ratio of banked area to developed area will cause a high cost of bios credits, leading to slow trading because of a bad deal for the developer who might as well use current regulations or do nothing.

Resolution: Resolution can be achieved through adaptive management, that is, the management authority for the market must check that the biodiversity requirements are accurate and not biased, audit conservation goals, conduct research to fill gaps in management capability, increase/decrease financial evaluation of biodiversity asset, increase efficiency of biodiversity management by strategic use of key/connective habitat, foraging productivity (more food), reproductive success (den sites, connectivity), decreased mortality (predator/disease/competitor management, fire/logging/roading).

Issue 2: Lack of confidence in the market has been an ongoing issue. Market participants are discouraged because there is no evidence that the process can demonstrate progress in managing biodiversity. What are the appropriate criteria to make decisions regarding success and failure? Uncertain economic value (market size and market quality) of biodiversity in any one region will preclude trading, or limit the market if trading does begin.

Resolution: Values can be clearly set by defining the important units of biodiversity that are relevant to the biodiversity banks in question, quantifying minimum requirements for successful management (hectares of EEC or hectares of high quality habitat for species or populations. The tools to make these definitions and quantifications are readily available through the systems developed in the RFA process and other, similar processes for setting criteria and defining targets (Joint Australia-New Zealand Environmental and Conservation Council Forest Reserve Criteria, NSW local environmental plans, catchment blueprints, C-Plan, state and national recovery plans). These minimum values and targets can be used on a comparative basis to set prices. For instance, a highly reserved tablelands forest ecosystem would not have the same value as a coastal forest ecosystem that is under-represented in the reserve system.

Issue 3: There are no clear means of either maintaining the market in perpetuity or achieving a permanent economic resolution or 'endgame' for the process of biodiversity banking.

Resolution: Demonstrate ongoing value in the biodiversity asset and enable perpetual trading. This can be done through monitoring economic contribution of the biodiversity asset. Payments from eco-tourism, wildlife use, compatible and sustainable enterprises within the bank (for instance, a commercial property already present on the land may contribute an annual rental or other income) or financial arrangement (investments of funds) can all contribute to an asset which increases over time and thus drives a market. The alternative is to plan an exit strategy based around investment of the funds acquired from the original sale of biodiversity credits. There are many legal vehicles which can provide the security and services required to deliver a financial return and an ongoing enterprise (art museums such as the Guggenheim may provide a good template for a successful endgame for biodiversity banks).

Issue 4: Inefficient delivery of services by the bank's administrators and management. That is, if the agency or institution tasked with delivering the operational structure of the bank is inefficient or incapable of delivering on requirements to handle investments and conduct transactions, businesses will not participate in the market.

Resolution: Ensure that staff and management of the administrative bodies are well trained, motivated to succeed and sufficiently resourced to carry out efficient delivery of the required services. If the biodiversity bank is operated by public servants on low pay, with little experience and wide responsibilities, it will not succeed. For instance, it is feasible to imagine a scenario where two junior technical staff members, trained at an urban university in ecology, are tasked with consideration of hundreds of biodiversity transaction requests. They are not motivated by pay, training or philosophic background to enable and manage a biodiversity trading scheme in the context of further development. However, if highly trained and experienced scientific officers and land management personnel are tasked with delivering a well resourced programme to achieve success in biodiversity management, based on clear goals and responsibility, the scenario would be very different. Monitoring, motivating and training biodiversity bank staff is essential to avoid problems arising from this.

Issue 5: The process is rendered unworkable due to scientific criticism based on prejudice and lack of experience with the management of social, economic and operational issues. If the prevalent attitude and belief among natural scientists, particularly 'practicing ecologists' is that biodiversity banks could not work to solve problems and are only a means of enabling further development, this can and will sway public opinion and thus market participation.

Resolution: Resolution can be achieved through communication and clarity, with regard to prejudice and belief that biodiversity banks will not work. In NSW, the regulations for bio banking will not allow any developments that are proscribed by the previous approval process, for instance, although many ecologists apparently believe that it will. The best resolution is through measuring success objectively and scientifically with regard to biodiversity goals, not with regard to preconceptions. For instance, a recent paper (Adam, P., pers. comm.) found that a recreated ECC, ten years after planting, was still significantly different ($P > 0.05$) from natural examples of the same communities. Professor Adam went on to report that, although the recreated habitat had great and new conservation values, it was nonetheless still not a success, as there was still a significant difference. I would contend that the exercise in question has had remarkable success in increasing the biodiversity on that particular patch of land, and the work done clearly demonstrates my point. The resolution comes from the continuing scientific measurement of the biodiversity asset (monitoring). In the first few years of trading, the biodiversity bank and market would be prudent to invest in frequent measurements of their assets (e.g. monitoring and field surveys).

Local resolution, global implications

In conclusion, let us imagine a successful resolution to some biodiversity issues in NSW. On the coast, Aboriginal land councils participating in bios banking schemes have an income stream associated with maintaining the biodiversity asset – in fact, they may have the management contract and be paid in dollars to deliver traditional land management services (put in the fire stick) or harvest bush tucker (native wild food) for further profit and enjoyment. In the tablelands, graziers participating in bios banking schemes could be receiving income from inaccessible blocks or rehabilitated woodlands on formerly unprofitable grazing paddocks. In the sheep–wheat belt, landowners with native grasslands or mallee on their properties might get income from their biodiversity asset, rather than from destroying it to put in another cropping paddock that will contribute to salinity problems. At the urban interface in Sydney, Newcastle and Wollongong, property owners with remnant native vegetation they have preserved might get premium prices for their Blue Chip biodiversity assets (and get to keep their property, too). And in the arid zone, a very high-value little digger, the marsupial mole, may provide much needed income to remote communities that currently do not know they possess a biodiversity gem.

By using the lessons from Aboriginal and early European land managers, we may be able to choose a wise course for land use. By using science – ecology and economics and engineering – we may find some solutions we didn't know about. By negotiating and communicating, we may be able to do so without polarization or conflict.

Throughout Australia the value of biodiversity is now truly recognized, and many initiatives such as the Bush Broker allow voluntary investment in this asset. There are valuable precedents particularly in the NSW biodiversity legislation and regulations, because they are a first in terms of regulated investment in the priceless asset of biodiversity.

Acknowledgement

The author thanks Justin Williams and Adam Fawcett for their excellent assistance in the preparation of this chapter.

Bibliography

Brand, David (2005) 'Session: Demand for forests and forest products 2020 Forests Woods and Livelihoods: Can the value of ecosystem services pay for the conservation of the world's remaining tropical rainforests?', in Brown, A. G. (ed) *Forest, Wood and Livelihood*. Record of a Conference Conducted by the Australian Academy of Technological Sciences and Engineering Crawford Fund, Parkville, Victoria

Clark, M. (1969) *A Short History of Australia*, Tudor Press, Balmain, Sydney

Department of Environment and Conservation (2005) 'Biodiversity certification and banking in coastal and growth areas', www.environment.nsw.gov.au/business/biobankscheme.htm, accessed in October 2005

Department of Environment and Conservation (2006) 'Questions and answers about the biodiversity banking and offsets scheme', www.environment.nsw.gov.au/threatspec/biobankscheme.htm, accessed in October 2006

Gibbons, P., Briggs, S. V. and Shields, J. M. (2002) 'Are economic instruments the saviour for biodiversity on private land?' *Pacific Conservation Biology*, vol 7, pp223–228

North, A. J. (1911) '*Nests and Eggs of Birds Found Breeding in Australia and Tasmania*', Volume 3, Australian Museum, Sydney

Pressey, R. L., Hager, T. C., Ryan K. M., Schwarz, J., Wall, S., Ferrier, S. and Creaser P. M. (2000) 'Using abiotic data for conservation assessments over extensive regions: Quantitative methods applied across New South Wales, Australia', *Biological Conservation*, vol 96, no 11, pp55–82

Shields, J. M. (1997) 'Biodiversity credits: A system of economic rewards for sustainable ecological developments', Invited Paper, University of New England, Armidale, NSW

Shields, J. M. (2003) 'Wildlife management in NSW native forests', in Lunney, D. (ed) *Conservation of Australia's Forest Fauna* (2nd edn), Royals Zoological Society of NSW, Sydney 2004

Part V

Conclusion

15

The Future of Biodiversity Offset Banking

Nathaniel Carroll, Ricardo Bayon and Jessica Fox

In closing, let us leave you with a few thoughts for the future. The following are big and sometimes incipient ideas, but they appear to be some of the trends and undercurrents now shaping the future of biodiversity offset banking.

First, it has been said before, but it bears repeating: mitigation (or offsetting) should never be the first action taken when conserving biodiversity in a land development setting. Indeed, it should be the last step in a so-called 'mitigation hierarchy', where the first order of business is to go to great lengths to avoid as many of the harmful impacts to biodiversity as possible. Only once it has been determined that the impacts are inevitable, should the project be undertaken. The second step is to minimize or reduce the extent of these impacts. Once the damage has been minimized, then – and only then – should the third step (mitigation or offsetting) come in. This book is only about one possible tool (biodiversity banking) aimed at that third step in the hierarchy. For this tool to be effective, however, the first two steps must be diligently and fully executed.

Second, as biodiversity banking develops, it is becoming increasingly clear that the process cannot be looked at in isolation. Like any piece of a larger system, biodiversity banking will be most effective when built into planning at a much broader spatial and ecological scale. Otherwise, we are at risk of winning the environmental battle but losing the war. We need to ask what is the endgame for any given landscape? What sort of landscape can we live with – environmentally, economically and socially? And, perhaps most important: How do we get there? What other ecological processes and markets are involved? Atmospheric, hydrological, geologic, biotic, human? Species and their habitat depend on and interact with factors that are often outside the scope of a specific regulator's jurisdiction. This expansion of scope will require inter-agency coordination, new market development and considerable foresight, among others. USFWS is taking the first steps towards this sort of broader vision with the signature of an inter-agency agreement to consider habitat banking for species other than those listed on the endangered species list (NRCS, 2007). And, as we saw in Chapter 13, the Australian state of New South Wales is a step ahead in broadening their mitigation market past just endangered species habitat to encompass all terrestrial biodiversity.

As the vision broadens, and as these markets expand, it is important to realize that biodiversity markets will likely integrate more and more closely with other environmental markets (markets for carbon, water and wetlands). There is already considerable interest in this sort of 'bundling' from credit suppliers wanting to maximize profit from their assets. But, as was pointed out in Chapter 11, there are questions of additionality or double-dipping that will need to be resolved. Overall, however, this integration, or stacking, of ecosystem service markets has the potential to help further provide that essential 'value' to nature's goods and services. But it will only work to the extent that it truly contributes to the long-term preservation and conservation of environmental assets. If it is merely a way to squeeze more profit from a piece of land, with little or no regard for its ultimate environmental impact, then it is doomed to failure.

Third, as the market grows and more players enter the field, quality becomes more and more important. This means that verified and enforced standards for ecological performance are needed to make sure that the market remains credible and robust (Chapter 7). As we have seen in most other environmental markets (and the carbon markets in particular), the future of these schemes rests on their ability to prove that they are making an environmental difference. If good species habitat is being destroyed in exchange for non-functioning ecosystems and 'bad' species habitat, then the system is a failure and it will, ultimately, not be allowed to continue. Quality matters as much (or more) in terms of biodiversity banking as it does in any other business. Indeed, perhaps the time has come to establish agreed-upon (and enforced) quality standards in terms of biodiversity banking.

Fourth, it is important to note that biodiversity banking works best when it goes beyond the status quo and when it is used not simply to prevent a loss to biodiversity, but rather when it is used to ensure a net gain for biodiversity. In the US, this has been the subject of some thought and debate. Witness that in some cases, when a unit of biodiversity is impacted, twice, triple or more the number of biodiversity units are required to be protected or restored as compensation. In this way, habitat loss can be reversed, not just stabilized.

Finally, conservation banking is beginning to feel a surge of capital investment. There is an emerging recognition of natural resource stewardship and restoration as a dynamic area for investment around the world. The global carbon markets are now transacting some $30 billion each year, and that has piqued many investors' interest. They are beginning to look at mitigation banking, forestry carbon and water-quality trading alongside renewable energy, green building, conservation development and water-related green tech as sectors worthy of increased interest. Parthenon Capital's investment in Wildlands is but one of the first such entries into this growing field. Investors trying to ensure that they do not miss out on the 'next carbon' are beginning to place bets on biodiversity banking.

With this greater scrutiny and interest should come greater responsibility. As the field broadens, it will become increasingly important for it to become as transparent, and as quality-oriented, as possible. Only then will this sort of interest from capital markets be sustained and not squandered.

Clearly this book comes at a time of mounting interest in market-based approaches from regulators, bankers, credit buyers, investors and the environmental community. We hope this book will serve as a foundation of knowledge from which industry players will build and improve. We believe complete and honest information leads to stronger and better markets. And we feel that the market for biodiversity mitigation can help give conservation a better say when it comes time to make economic decisions.

We said it before, and we'll say it again: Species banking and biodiversity offsets may be the worst of all possible systems (and we hope this book will help make them better), but they are better than all the others, and they are certainly better than the alternative: a model that sees biodiversity as something that is virtually worthless. Let us therefore look ahead to a future where the California tiger salamander is not only treasured and protected, but is also recognized on the balance sheets of businesses. And our natural infrastructure, species and ecosystems, are valued as clearly as our built infrastructure, highways and housing developments.

Reference

Natural Resources Conservation Service (2007) 'USDA, DOI, and AFWA sign habitat credit trading agreement', www.nrcs.usda.gov/NEWS/releases/2007/habitatcredittradingagreement.html, accessed in August 2007

Appendix I

US Federal Guidance for Conservation Banks

Guidance for the Establishment, Use, and Operation of Conservation Banks

United States Fish and Wildlife Service,

Department of the Interior

2 May 2003

Introduction

Purpose and scope of guidance

This document provides guidance on the establishment, use and operation of conservation banks for the purpose of providing a tool for mitigating adverse impacts to species listed as threatened or endangered under the Endangered Species Act of 1973, as amended. This guidance can also be used to aid in the establishment of banks for candidate species. The Service envisions that banks will mainly be used for candidates in conjunction with Candidate Conservation Agreements with Assurances or as a precursor to a multiple species Habitat Conservation Plan (HCP) effort that covers listed and non-listed species.

The policies and procedures discussed herein are applicable to the establishment, use and operation of public conservation banks, privately sponsored conservation banks and third party banks (i.e. entrepreneurial banks). The guidance they provide is intended to help Service personnel: (1) evaluate the use of conservation banks to meet the conservation needs of listed species; (2) fulfill the purposes of the ESA; and (3) provide consistency and predictability in the establishment, use and operation of conservation banks. In this regard, it is important to apply consistent standards and principles of mitigation whether mitigating through conservation banks or through other means. The purpose of this policy is not to set the bar higher for conservation banks than for other forms of mitigation, but articulate generally applicable mitigation standards and principles and to explain how they are to be accomplished in the special context of conservation banks.

Conservation banks are a flexible means of meeting a variety of conservation needs of listed species. The use of conservation banks should be evaluated in the context of unavoidable impacts of proposed projects to listed species. In some cases, the use of off-site banks may be the only mitigation option when on-site conservation measures are not practicable for a project or when the use of the bank is environmentally preferable to on-site measures. In general, no two conservation banks will be used or developed in an identical fashion. However, as demand for conservation banking increases, it is important that the essential components and operational criteria of conservation banks are standardized to ensure national consistency.

Background

Conservation banking is attractive to landowners and land managers because it allows conservation to be implemented within a market framework, where habitat for listed species is treated as a benefit rather than a liability. From the Service's perspective, conservation banking reduces the piecemeal approach to conservation efforts that can result from individual projects by establishing larger reserves and enhancing habitat connectivity. From a project applicant's perspective, it saves time and money by identifying pre-approved conservation areas, identifying 'willing sellers', increasing flexibility in meeting their conservation needs and simplifying the regulatory compliance process and associated paperwork. From the landowner's perspective, it provides a benefit, an opportunity to generate income from what may have previously been considered a liability.

Directing smaller individual mitigation actions into a bank streamlines compliance for the individual permit applicants or project proponents while providing a higher benefit to the natural resources. Banking allows a collaboration of private/public partnerships to maintain lands as open space, providing for the conservation of endangered species. Local communities as a whole benefit by being assured that their natural resources will be protected and open space maintained.

Conservation banking can bring together financial resources, planning, and scientific expertise not practicable for smaller conservation actions. By encouraging collaborative efforts, it becomes possible to take advantage of economies of scale (both financial and biological), funding sources, and management, scientific and planning resources that are not typically available at the individual project level.

What is a conservation bank?

A conservation bank is a parcel of land containing natural resource values that are conserved and managed in perpetuity, through a conservation easement held by an entity responsible for enforcing the terms of the easement for specified listed species, and used to offset impacts occurring elsewhere to the same resource values on non-bank lands. Bank parcels are typically large enough to accommodate the mitigation of multiple projects. A project proponent will secure a certain amount of natural resource values within the bank to offset the impacts to those same values off-site. The bank is specifically managed and protected by the banker or designee for the natural resource values. The values of the natural resources are translated into quantified 'credits'. Typically, the credit price will include funding for the long-term natural resource management and protection of those values. Project proponents are, therefore, able to complete their conservation needs through a one-time purchase of credits from the conservation bank. This allows 'one-stop-shopping' for the project proponent, providing conservation and management for listed species in one simplified transaction.

A bank can be created in a number of different ways: (1) acquisition of existing habitat; (2) protection of existing habitat through conservation easements; (3) restoration or enhancements of disturbed habitat; (4) creation of new habitat in some situations; and (5) prescriptive management of habitats for specified biological characteristics. Banks can be created in association with specific projects, or can proceed from a circumstance where a project proponent sets aside more area than is needed for the immediate project, and is willing to protect the remaining area and thus generate credits, or where the specific project is implemented over a longer period of time. A conservation bank can also be created as an entrepreneurial effort in anticipation of an independent customer base with a number of different potential projects.

Once conservation banks are established, each credit they sell is considered to be part of the environmental baseline. As a result, future project evaluations and listing or delisting decisions can be made in a more stable ecological context. This stability is one of conservation banking's greatest assets, from both ecological and economic standpoints. For this reason, it is particularly important that conservation banks be established in perpetuity, regardless of the future status of the species for which the bank was initially established.

Wetland mitigation banking vs conservation banking

The wetland mitigation banking policy was finalized in November of 1995 (60 FR 58605). The main concept behind wetland mitigation banking is similar to that of conservation banking: to provide compensation for adverse impacts to wetlands and other aquatic resources in advance of the impact. Under the guidelines established for section 10 of the Rivers and Harbors Act and section 404 of the Clean Water Act, impacts to wetlands are mitigated sequentially by avoiding impacts, minimizing impacts, and then, as a last resort, compensating for those impacts. Compensatory mitigation involves creating, restoring or enhancing lost function and values of the wetlands. In the absence of mitigation banking, this often led to small, isolated wetlands being restored without long-term value. Wetland mitigation banking was used to consolidate smaller mitigation requirements for wetland impacts. Typically, the mitigation bank policy focused on establishing credits based on the restored or enhanced value of the area, and discouraged the establishment of 'preservation' banks. This makes sense when the functions of wetlands on the landscape are considered in the context of a no net loss policy.

Conservation banking transferred the concept of wetland mitigation banking into endangered and threatened species conservation with a few slight differences. While in wetland mitigation banking the goal is to replace the exact function and values of the specific wetland habitats that will be adversely affected by a proposed project, in conservation banking the goal is to offset adverse impacts to a species. These different goals account for differences in the policies guiding operations of the two banks. In contrast to mitigation banks, an appropriate function of conservation banks is the preservation of existing habitat with long-term conservation value to mitigate loss of other isolated and fragmented habitat that has no long-term value to the species. It forces the Service to evaluate all issues surrounding banking in the context of the benefit to the species – a sharply contrasting standard to that of wetland banking, where the focus of mitigation is on maintaining function and values present in a particular watershed.

Endangered species conservation banking has been implemented in California since 1995, where the Service has worked with the State of California Department of Fish and Game (CDFG). The CDFG policy on conservation banking describes conservation banks as:

> *A conservation bank is privately or publicly owned land managed for its natural resource values. For example, in order to satisfy the legal requirement for mitigation of environmental impacts from a development, a landowner can buy credits from a conservation bank, or in the case of wetlands, a mitigation bank. Conservation banking legally links the owner of the bank and resource agencies, such as the Department of Fish and Game or the US Fish and Wildlife Service.*

Policy considerations

The Service's intent is that this guidance be applied to conservation bank proposals submitted for approval on or after the effective date of this guidance and to those in early stages of planning or development. We do not intend for the policy to be retroactive for banks that have already received agency approval. While we recognize that individual conservation banking proposals may vary, our intent for this guidance is that the fundamental concepts be applicable to future conservation banks.

Conservation banking can assist both the section 7 and section 10 processes in reaching their goals. Many activities authorized under these processes result in adverse effects to listed species, including habitat loss or modification. One way to offset these types of impacts is to include in the project design a plan that involves the restoration and/or protection of similar habitat on- and/or off-site. Purchasing credits in conservation banks is one method of protecting habitat off-site or on-site.

Authorities

Section 7

Section 7(a)(1) of the ESA requires that all federal agencies ... in consultation with and with the assistance of the [Service], utilize their authorities in furtherance of the purposes of [the ESA] by carrying out programs for the conservation of [listed species]. Section 7(a)(2) of the ESA also requires each federal agency to consult with the Service regarding effects of their actions to insure that the continued existence of listed species will not be jeopardized and that designated critical habitat will not be destroyed or adversely modified. Impacts to listed species are minimized by including conservation measures for the listed species in the federal agency's project description. These conservation measures could include, if appropriate, protection of off-site listed species habitat through purchase of credits in a conservation bank.

Section 10

Section 10(a)(1)(B) of the ESA authorizes the Service to issue to non-federal entities a permit for the incidental take of endangered and threatened species. This permit allows a non-federal landowner to proceed with an activity that is legal in all other respects, but that results in the incidental taking of a listed species. A HCP must accompany an application for an incidental take permit. The purpose of the HCP is to ensure that the effects of the permitted action on covered species are adequately minimized and mitigated and that the action does not appreciably reduce the survival and recovery of the species. Mitigation may include off-site protection of the listed species and its habitat and may take the form of purchasing credits in an approved conservation bank. Credits must be acquired by the permittee prior to commencement of actions authorized by an incidental take permit and intended to be mitigated by those credits.

Planning considerations

Goals and objectives

The overall goal of any conservation bank should be to provide an economically effective process that provides options to landowners to offset the adverse effects of proposed projects to listed species. The goal of a bank should be focused on producing conservation benefits for the species for which the bank is being established. For instance, many species are facing the threat of habitat loss and fragmentation. By consolidating and managing the high-priority areas in a reserve network, the threat of fragmentation may be reduced and the species can be stabilized. The species recovery plan and conservation strategy are among the tools available to develop the goals and objectives for establishing conservation banks. The important point in establishing a bank is to site banks in appropriate areas that can reduce the threat of fragmentation and provide management measures that address other threats that a species might encounter, such as cowbird parasitism, non-native invasion or disruption of natural disturbance regimes.

Conservation strategy

Any conservation strategy that the Service develops should identify threats, conservation needs and actions that address those threats and needs in the service area. This information can then help the Service evaluate whether the banking concept, the geographic location, the size, and management for the species is appropriate. The recovery plan can help guide the Service in evaluating whether creation of a bank will contribute to the conservation needs of the species. However, in instances where the recovery plan is not specific, is not available or is outdated, the Service may consider options to assess bank effectiveness. One option is to develop a local step-down approach or strategy to addressing the needs of the species.

The conservation strategy or species conservation needs should address the factors which caused the species to be listed and must be based on sound scientific principles. The main threat to a majority of the listed species is habitat loss and fragmentation of the remaining habitat. To reduce this threat, conservation biology principles have often been used to conserve populations of species in a reserve network, consisting of core populations that are interconnected by dispersal corridors. Conservation banking can aid in such a strategy by adding conservation areas that are permanently managed to the reserve network.

Principles of conservation bank evaluation

Both section 7 and section 10 require the evaluation of a project's adverse effects to a species and determine whether the proposed project, together with any offsetting measures, will jeopardize the continued existence of the species. The adverse effects and offsetting measures are evaluated in the context of the current status of the species and the threats to the species. Implicit in the approval of a conservation bank is the recognition that adverse effects to a species may be offset by the conservation improvements offered by the bank. The Service is agreeing that projects which include adequate mitigation of impacts through the purchase of bank credits are consistent with the conservation needs of the species covered by the bank.

For the Service to determine whether to approve a proposed bank, the Service should determine whether the bank will provide adequate mitigation for the species. When the Service evaluates a proposed mitigation package that is intended to offset adverse effects to listed species, the Service evaluates whether the mitigation will fit within the conservation needs of the species.

For instance, if a proposed project involved habitat loss, the offsetting measure may be to conserve habitat in a location that contributes to the overall conservation strategy of the species, which may be located in a corridor or core area that supports essential breeding habitat. The conservation bank will provide mitigation to offset impacts and therefore should be evaluated in the same fashion. The best way to justify approving a bank is to evaluate whether the bank fits into the overall conservation needs of the listed species the bank intends to cover.

Two issues of paramount importance in evaluating any conservation bank are the siting of the bank and its management program. Although recovery plans for individual species will rarely, if ever, identify particular parcels as desirable sites for conservation banks or other conservation actions, they often identify broader areas within which recovery efforts will be focused. Conservation banks sited in these areas can create mitigation opportunities that both increase the options available to regulated interests and contribute to the conservation of the species. For species without recovery plans, or with plans that do not clearly identify those areas where recovery efforts will be primarily focused, conferral with the Service is especially important, to identify those areas it regards as of particular value in conserving the species.

For many species, individual conservation banks are seldom large enough, by themselves, to support a viable population of a threatened or endangered species over the long term. But if the bank is located next to an existing area managed for the conservation of that species, even a small conservation bank may increase the likelihood that a viable population can be maintained there. Similarly, if a bank is sited to encourage dispersal between two areas managed for the conservation of the species, the bank may increase the likelihood of the species surviving at both locations and thus provide a benefit proportionally larger than its actual area. In some instances, banks may be able to provide replacement habitat for species currently occupying nearby unmanaged habitats at risk of becoming unsuitable because of succession. Sites that otherwise appear to be good locations for conservation banks may turn out, on closer examination, to be inappropriate because of anticipated land-use changes in the surrounding area. These and other considerations relevant to the siting of a conservation bank should be taken into account at the outset and discussed with the would-be banker to ensure that needs for species conservation is compatible with the banker's objectives.

No less important than siting is the bank's management program. Seldom will the needs of a threatened or endangered species be met on a completely unmanaged piece of property. More commonly, an active management program – to control invasive exotic species, replicate natural disturbance regimes, prevent an area's use by off-road vehicles, illegal garbage dumpers or others, and address myriad other threats – is essential to ensure that the potential conservation value of a particular property is realized and maintained. These management needs should be anticipated and provided.

Eligible lands

Conservation banks may be established on tribal, local, private, or state lands where managing agencies maintain or will maintain habitat in the future. Use of conservation banks on federal lands is not precluded under this guidance, although there may be special considerations concerning applicability of conservation banks on federal lands. Therefore, future guidance will be forthcoming on this point. Until such time, use of conservation banks on federal lands would occur only on a case-by-case basis after review and approval by the Director.

Land used to establish conservation banks must not be previously designated for conservation purposes (e.g. parks, green spaces, municipal watershed lands), unless the proposed designation as a bank would add additional conservation benefit. For instance, it may be advantageous to place in a conservation bank the biological and habitat benefits that a species has gained under a Safe Harbor Agreement, where the landowner would agree to maintain those resource values in perpetuity.

Where conservation values have already been permanently protected or restored under other federal, state, tribal, or local programs benefiting federally listed species, the Service will not recommend, support or advocate the use of such lands as conservation banks for mitigating impacts to species listed under the ESA. This includes programs that compensate landowners who permanently protect or restore habitat for federally listed species on private agricultural lands, as well as easement areas associated with inventory and debt restructure properties, lands protected or restored for conservation purposes under fee title transfers, lands protected by a habitat management agreement (unless the agreement is extended in perpetuity by a bank agreement) or habitats protected by similar programs. For example, lands conserved under the section 6 HCP land acquisition grant program would not be available for conservation bank establishment. Where federal funds have been used in the establishment of a bank, the allocation of credits to the bank will be proportionate to the non-federal contribution. A bank capable of sustaining ten credits, but with a 50 per cent federal contribution, will be allocated five credits.

Site selection

The Service will give careful consideration to the ecological suitability of a site for achieving mitigation. The Service will evaluate the location, size and configuration of the proposed bank. Additional items to consider when determining the suitability of an area as a conservation bank might be topographic features, habitat quality, compatibility of existing and future land-use activities surrounding the bank, and species use of the area.

Conservation biology principles suggest that conserving large, unfragmented habitat blocks, to reduce the edge effect, in a reserve network will help to maintain viable populations. A conservation bank could be large enough to maintain a viable population within its boundaries or be situated in a strategic location that would add to an already established conserved area. The conserved area might be a privately owned mitigation site established under a HCP, or a state park. Banks could also be sited between two larger areas in a corridor that will maintain connectivity for dispersing individuals.

Bank boundaries should ordinarily be drawn so as to exclude developed areas or other areas that cannot reasonably be restored. Potential banks that encompass such areas should only be approved if the activities that will occur on these areas will not impact the value of the bank for conservation or if the resulting value will be sufficient to warrant conservation in spite of the developed areas. However, if the latter is the case, we must have the assurance that the impacts will not change over time in a manner that will decrease the value of the bank. Factors to consider include, but are not limited to, activities that may result in incidental take, habitat degradation and contamination.

It is also possible to establish conservation banks within the boundaries of a proposed project, such as a HCP planning area, if it is both feasible and appropriate given the habitat type and species needs. If the project plan area contains sufficient land and the project impacts are fairly localized, it may be possible, or even desirable, to designate a conservation bank within its boundaries. Ultimately, the credits purchased from a conservation bank must provide biologically comparable habitat to the area affected by the activity to be mitigated.

Inclusion of buffer area

In general, it is important that banks be of sufficient size to ensure the maintenance of ecological integrity in perpetuity. However, the minimum or maximum sizes of parcels of land designated as a conservation bank will be determined on a case-by-case basis depending on the needs of the species proposed to be covered in the bank, the location of the bank and the habitat values that are provided. Bank boundaries must encompass all areas that are necessary to maintain the habitat function specific to the species covered by the bank, which may include the appropriate buffer against edge effects from adjacent land use.

These buffer areas may not always consist of habitat that is necessary for the species included in the bank. However, limited credits may be given for the inclusion of these buffer areas only to the degree that such features increase the overall ecological functioning of the bank.

Role of restoration, enhancement and creation of habitat

Conservation banks will rely on a range of strategies to achieve and maintain mitigation in perpetuity on existing functioning and occupied habitat for a majority of those species facing threats of habitat loss and fragmentation. Such strategies include preservation, management, restoration of degraded habitat, connecting of separated habitats, buffering of already protected areas, creation of habitat and other appropriate actions. The preservation strategy will be employed for those species in which the habitat is not easily restored or created, or the information on how to accomplish the restoration or creation of habitat is either not known or unreliable. Other species may rely heavily on creation or restoration of habitat as part of a conservation bank. The reliance on restoration, enhancement or creation of habitat as part of a bank strategy will be species specific. All conservation banks must have an element of management that will maintain the habitat for the species in the bank.

Conservation banks can be used in instances where significant restoration, enhancement or creation of habitat are necessary. However, an appropriate credit system will need to be developed to address these situations. If restoration is proposed as part of the conservation bank, appropriate measures should be implemented to increase the likelihood of success. One way to increase the likelihood of success is to require some method of ensuring performance, such as authorizing sale of credits only upon completion and verification of restoration outcomes.

One strategy is to designate preservation credits for the protection of existing habitat and restoration credits for the restoration, enhancement and preservation of areas not currently providing suitable habitat. The need for this type of distinction will vary depending on the specific ecological situation and the conservation strategy being employed. For example, we may determine that a species cannot afford any reduction of its total available habitat. For this reason, we may require the development of a process that provides for one acre to be protected and one acre to be restored for every acre of habitat destroyed. Taken to its full extent, this conservation strategy would result in half of the existing habitat being protected with the remaining habitat being replaced through habitat restoration.

Criteria for use of a conservation bank

Project applicability

Activities regulated under section 7 or section 10 of the ESA may be eligible to use a conservation bank, if the adverse impacts to the species from the particular project are offset by buying credits created and sold by the bank. Credits from a conservation bank may also be used to compensate for environmental impacts authorized under other programs (e.g. state or local regulatory programs, transportation projects, NEPA or state equivalent). In no case may the same credits be used to compensate for more than one activity; however, the same credits may be used to compensate for an activity that requires authorization under more than one program. In other words, once a credit is sold to offset an adverse impact, that same credit cannot be sold again.

Service area

In general, the service area of a conservation bank is identified in the bank agreement and defines the area (e.g. recovery unit, watershed, county) in which the bank's credits may be used to offset project impacts. In other words, if proposed projects fall within a specific conservation bank's service area, then the proponents of those projects may offset their impacts, with the Service's approval, by purchasing the appropriate number of conservation credits from that bank. In the event that the proposed projects fall within the service area of more than one conservation bank, then the project proponents would have the option of using any of the banks or perhaps even more than one bank.

Designation of the service area should be based on the conservation needs of the species being conserved. For this reason, banks generally should be located within areas designated in recovery plans as recovery units or other applicable recovery focal area, and their service areas should correspond to the recovery areas in which they are located. If there is no applicable recovery plan, banks should be sited, and service areas should be designated, to serve a comparable purpose.

Two exceptions to the preceding general guidance should be noted. First, some projects may be located outside a recovery unit. Banks located within recovery units should be able to provide credits for such projects. In such situations, the project to be mitigated will have little or no detrimental impact on recovery prospects, and the mitigation bank will aid those prospects.

A second exception to the general guidance regarding service areas concerns projects located in recovery units and undertaken after the recovery objectives for those areas have been achieved. Such projects should be able to buy mitigation credits from banks located in other recovery units. Allowing such projects to do so will help achieve the recovery objectives in the recovery unit where the bank is located, without hurting these objectives in the area of the project requiring mitigation.

The service area is an important component for bank owners, who will need to evaluate the marketability of their banks (i.e. the potential demand for their conservation credits). The individual bank owner has the responsibility to determine if a bank will be profitable. The bank agreement should clearly define any constraints that are found within the service area. These might include exclusion of areas that are key to a regional reserve system, such as projects that occur within corridors or core reserve areas. Or, a particular bank in a county could have a service area corresponding to the regional plan boundary, yet limit projects using the bank to those that are in fragmented, isolated, highly urbanized areas not contributing to the regional reserve system.

Credit system

Credits are the quantification of a species' or habitat's conservation values within a bank. The conservation values secured by a bank are converted into a fixed number of credits that may be bought, sold or traded for the purposes of offsetting the impacts of private, state, local or federal activities. In its simplest form, one credit will equal one acre of habitat or the area supporting one nest site or family group. Credit values are based upon a number of biological criteria and may vary by habitat types or management activities. When determining credit values, some of the biological criteria that may be considered include habitat quality, habitat quantity, species covered, conservation benefits, including contribution to regional conservation efforts, property location and configuration, and available or prospective resource values.

In general, the credit system for a conservation bank should be expressed and measured in the same manner as the impacts of the development projects that will utilize the bank. For instance, if a development project will permanently remove some amount of habitat acreage and a number of pairs of a species, then the bank's credits should be expressed in terms of acreage and pairs. If effects are evaluated in terms of losses of family groups due to timber activities, then the bank credits should be established in terms of the number of family groups being conserved. The method of calculating bank credits should be the same as calculating project impact debits.

In some instances a bank may contain habitat that is suitable for multiple listed species. When this occurs, it is important to establish how the credits will be divided. For instance, once a project buys a credit for one species, that credit cannot be sold again for another species. If the proposed project impacts multiple species and the bank contains the same multiple species, then the credits can be sold for in-kind replacement. As a general rule, overlapping multiple species credits can overlap for a single project, but not multiple projects.

If the bank is a preservation bank, the credits should be based on the biological values of the bank at the time the bank agreement is established. Because some populations may vary in size due to natural dynamics, an agreement should be made, before the bank agreement is finalized, as to the number of credits in the bank, especially if the credits are based on the number of individuals or nesting pairs. This is a risk both for the Service and the banker. The risk to the Service is that the credit overestimates the average populations of the bank. The risk for the banker is that the agreement could be made in a low population year, depressing the amount of credits that the bank could have received. A study might be undertaken to determine the average populations occupying the bank, but this would be time-consuming and expensive for the banker and the Service.

An alternative would be to use incentives to arrive at a fair accounting for both the banker and the Service. An initial allocation of credits could be made to the bank based on the best available information on species average population sizes. This number would be set on the low end of the spectrum. Additional credits would then be awarded to the banker based on subsequent performance. When mutually agreed-upon mitigation outcomes or conservation milestones (i.e. the standards that must be met in order to earn credits above the initial allocation) are attained, the Service would authorize the additional credits.

At the time that the first credit in a bank or phase of a bank is sold, the land within the bank or its phase must be permanently protected through fee title or a conservation easement, with any land-use restrictions set in perpetuity for the land legally established. Consequently, once any credit in a given bank or phase is sold, the entire area is automatically and legally protected, regardless if the rest of the credits in the bank or phase are sold, thereby eliminating future fragmentation of habitat.

Every conservation banking agreement (CBA) should specify the methods for determining credits within the bank and debits outside the bank, setting performance standards to calculate credit availability, and devising accounting procedures to track the creation and use of such credits. If several conservation banks are created for the same species, the Service will use a consistent methodology for determining credits in each of them and make that methodology publicly available. That methodology should also be consistent with the methodology used to determine mitigation requirements for activities mitigated by means other than the purchase of credits from conservation banks.

Credits associated with a mitigation activity (as well as debits associated with an activity requiring mitigation) should reflect an assessment of the degree of beneficial (or detrimental) impact of the activity on the prospects for the affected species' survival. In theory, population viability analyses could be used to quantify the degree of impact on survival prospects. In practice, however, the information needed for rigorous population viability analyses is often unavailable. As a result, the units of currency may take the form of surrogates for the extent of impact on population viability, such as occupied acres or nesting pairs beneficially or detrimentally affected. In determining credits or debits, the same types of activities may be weighted differently depending on where they occur (e.g. nearby or far from existing protected areas), or other factors (e.g. quality of habitat at the affected site). The rationale for any differential weighting schemes should be clearly articulated in the mitigation agreement or elsewhere.

Phased establishment

Conservation banks may be divided into sub-areas and implemented in phases. This approach is useful and appropriate in many circumstances. A prospective bank manager may not be sure there will be sufficient demand to use all of the potential credits. Therefore, the banker may decide to implement a conservation bank on only a portion of the habitat area during the first phase of the bank. Later phases of the bank would be added if and when the credits from this first phase are exhausted. Other situations justifying a phased approach include those in which a potential banker can only afford to enhance or manage a portion of the entire habitat area until revenue from the first phase is received, or when a potential project proponent is uncertain about the level of impact he or she will be creating over time and thus is uncertain how many conservation credits will be required.

Alternatively, the Service may want to seek the implementation of a bank in a phased manner. For example, in a situation where there is uncertainty regarding the level of future biological need within a specific area, it may be desirable to implement a process in which high-quality habitat receives priority designation for protection, and lands of lesser quality habitat or lands targeted for ecological restoration or enhancement activities would be designated for secondary phase protection. This would increase the likelihood of protecting habitat of the greatest ecological value, with habitat of lesser ecological value being protected only if needed.

A non-phased approach with a similar outcome would be to use weighted credits. Preservation of an acre of high-quality habitat might earn one credit, while preservation of an acre of low-quality habitat might earn half a credit. This would eliminate the need to prioritize land types for mitigation purposes. So long as the credit and debit methodology ensures that adverse impacts are fully compensated by corresponding beneficial actions of banks, it will not matter whether the first phase of a bank is high-quality or low-quality habitat. As a general rule, if the differences in habitat quality are sufficient to justify prioritization, then they are also sufficient to justify weighted credit valuations.

If a phased approach is to be taken, each phase must be evaluated on the assumption that its conservation value can stand on its own in the event that the additional phases are not added to the conservation bank in the future. For instance, if the species conservation strategy identifies the need for conservation areas to be established with a minimum size of 200 acres for the species population to be viable and the first phase of the bank is proposed for only 100 acres, then the Service may not want to approve the proposed phasing structure.

Relationship of the bank to the mitigation requirements

The most important consideration for any mitigation requirements – irrespective of variation between species and site specificity – is that they should be proportional to the extent of the impact and consistent from project to project. Mitigation requirements for individual projects may or may not be compatible with use of conservation banks. For example, the most appropriate mitigation for a particular project may involve emphasizing on-site preservation or restoration due to important local functions such as habitat protection for a species with a limited geographic range. There may be circumstances warranting a combination of on-site and off-site conservation measures, and, in these circumstances, conservation banks could be a useful tool. Conservation banks will only be available for use by projects that affect a species covered by the bank. In general, a bank established to provide credits for one group of species cannot be used to offset impacts to a species not part of the group, unless the Service establishes that the bank can provide the necessary conservation values to additional species, and implements the legal instruments to effect the change. The Service will approve the use of the conservation bank and establish the number and type of credits to offset impacts from a particular project.

In many situations, mitigation ratios are used to establish the amount of credits that will need to be purchased. While use of ratios may be based initially on a general knowledge of the relationship between the amount of habitat remaining and what should be conserved to achieve the site-specific conservation strategy, every adverse impact will need to be evaluated individually. In some circumstances, the ratios can be based on qualitative factors such as scale of impact or quality of habitat. This allows different ratios to be applied to ensure mitigation proportional to the impact. For example, a project involving loss of habitat that is small in magnitude and low in quality due to isolation might be expected to mitigate at a ratio of 1:2 (one bank acre to two project acres), while a project with a large area in high quality habitat might be expected to mitigate at a ratio of 2:1 (two bank acres to one project acre). Any mitigation ratio used, regardless of whether the ratio is greater than, less than or equal to 1:1, must be based on sound biological rationale that is easily explained, readily understood and consistently applied by the Service.

Coordination with other levels of government

Conservation banks covered by this policy are those established to meet the requirements of the ESA. State or local laws may also impose requirements that can be met by the measures provided for in a conservation bank. When that is the case, the Service requires that the relevant state or local government entity be given an opportunity to participate in the development of a CBA and to become a party to it. The Service will coordinate its requirements with those of state or local government entities to the extent possible in order to minimize expenses, burdens or duplicative requirements for bank sponsors, project proponents and other governmental agencies. Although the Service will encourage the appropriate state and local governmental agencies to participate in the development of CBAs and to become parties to them, the failure of such other agencies to participate in developing, or to sign an agreement that otherwise meets the requirements of this policy and of the ESA, shall not preclude the Service from entering into such an agreement. Any state and local agencies that participate in the bank agreement should be part of the conservation bank review team (CBRT) established to monitor the establishment, use and operation of the conservation bank

Public review and comment

The bank credits will be sold in conjunction with incidental take of listed species exempted under section 7 or authorized under section 10 of the ESA. Both of these processes have opportunities for public review. Section 7 consultations are conducted when federal agencies propose projects that have adverse effects to listed species. The federal action agencies are required to consider reasonable alternatives and analyze those impacts through the National Environmental Policy Act, which includes public review of the project, including mitigating factors. Through the section 10 process, all applications for permits authorizing the taking of listed species must be noticed by the Service for at least a 30-day public comment period. The use of credits from an established bank to mitigate actions in a HCP will require a permit application, notice and opportunity for public comment.

If approving the bank agreement is controversial, the Service may want to publish in the Federal Register advance notice of its intent to do so and invite public comment on the proposed agreement. If there are significant public concerns about the design or operation of a conservation bank, it is better to discover them before approving a banking agreement than afterwards.

Long-term management and monitoring

Management

Incorporating management into the bank agreement is key to the bank's success. With few exceptions, listed species and their habitat cannot be conserved without management of the conservation property. An active management program may consist of halting and removing illegal trash dumping, preventing trespassing that might include off-road vehicle use, and/or imitating the natural disturbance regimes that might include prescribed burns. The ultimate goal for any management plan will consist of maintaining the habitat for the continued use by the listed species conserved on site.

The amount of credits earned by a bank and available for sale to service area projects for mitigation is implicitly contingent on the bank's exercise of appropriate management to safeguard in perpetuity the species or habitat conservation values upon which the credits are based. This may require a range of management practices and responses, including those customarily identified as adaptive management practices. The choice of management strategies and the responsibility for engaging them to meet bank goals reside with the bank sponsor. As a general rule, species or habitat conservation value outcomes (e.g. numbers of nesting pairs and family groups, or enhanced or created habitat), rather than the implementation actions that are causal to those outcomes and values, are the standards by which the Service will evaluate banks and authorize issuance and sale of mitigation credits. In cases of phased development, banks that perform and produce good results earn more credits, and banks that perform poorly and produce inferior results earn fewer credits. Such an outcome-based management framework provides a robust, market-driven incentive for bankers to engage appropriate management practices and to take all necessary action to safeguard the conservation values that constitute the bank's permanent capital. While conducting management activities on the bank, the bank owner should be cautious not to degrade the status of other sensitive species.

Management of conservation banking areas can also include other non-mitigation related activities which involve public access. If sound professional judgement is exercised in determining the compatibility of a particular use in a particular bank area, there is no reason to exclude the public from these areas. Exercise of commonsense consideration of the biological constraints, public safety and conflicts between uses and compliance can result in a property that satisfies the habitat requirements of the species protected, while providing enjoyment and education to the public. While each mitigation bank will have its own set of constraints, this guidance is intended to encourage public access where it is appropriate and does not impinge on the primary function of habitat preservation.

Monitoring

Monitoring is the responsibility of the conservation bank. The scope of the monitoring program should be commensurate with the scope of the conservation actions undertaken by the bank. Biological goals of the bank provide a framework for developing a monitoring program that measures progress toward meeting those goals. The appropriate protective measures and level of monitoring will vary by individual circumstance, and an effective monitoring program should be sufficiently flexible to allow modifications, if necessary, to obtain the appropriate information. Monitoring provisions to measure and assess habitat protection, restoration or creation activities should be included in the CBA. Those provisions will include components to: (1) evaluate compliance based on current levels of credit authorization; (2) determine if biological goals and objectives are being met; (3) provide feedback information for subsequent management changes and adaptations, including remedial actions if necessary; and (4) substantiate and authorize additional increases in bank credits resulting from habitat restoration or creation activities, including phase-in of additional bank lands.

The monitoring program will be conservation bank-specific and will be based on sound science. The monitoring methods and standards should be structured to compare the results from one reporting period to another period, or to compare different areas within the conservation bank. Monitoring should be conducted at time intervals appropriate to the bank's management strategy. Monitored units should reflect the units of measurement associated with the biological goals (e.g. if a biological goal is in terms of numbers of individuals, the monitoring program should measure the number of individuals). Standard survey or other previously established monitoring protocols should be used. Though the monitoring for each ecosystem and each situation may differ, some factors that may be important to monitor include vegetative growth, the presence of invasive species (both plant and animal), water quality and listed species presence. Although the specific methods used to gather necessary data may differ depending on the species and habitat types, monitoring programs should use a multi-species approach when appropriate. In summary, the monitoring measures must be clearly identified in the bank agreement and they should be commensurate with the conservation goals of the bank.

To determine the level of success and identify problems requiring remedial action, the bank sponsor is responsible for monitoring the conservation bank in accordance with monitoring provisions identified in the bank agreement, and approved by the Service. The parties to the agreement should establish a CBRT that oversees the establishment, use and operation of the conservation bank. Monitoring reports should be submitted to the CBRT in accordance with the terms specified in the bank agreement.

Remedial actions

Every CBA must include provisions for a dispute resolution process applicable if the owners of the conservation bank fail to meet their obligations under the CBA. The dispute resolution process must also provide a method for disposal of the property to a third party capable of continuing the management of the property for species protection in the event of the current owner's inability to continue the operation of the bank for any reason. If necessary, a bond equal to the present value of the management costs may be posted or some other mutually agreed form of surety may be used to ensure performance. The agreement must contain provisions for contingencies that a prudent man would plan for; however, not every single possible contingency need be addressed. The bank should not be held responsible for offsetting acts of nature that are unforeseen, or foreseeable but unpredictable, such as earthquakes, floods or fires.

The CBA will stipulate the general procedures for identifying, implementing and funding remedial measures at a bank in the event of unexpected contingencies (fires, floods, etc.), particularly after credits have been sold by the bank. Contingencies that occur prior to the sale of credits may result in the temporary suspension of the recognition of those credits, pending full or partial remedial action. These remedial measures will be based on both information in the monitoring reports and the Service's on-site inspections. The Service, in consultation with the bank sponsor, will decide on the need for remediation.

Funding assurances

The bank agreement must identify and include a requirement for adequate funding to provide for the conservation bank's perpetual operation, management, monitoring and documentation costs. Therefore, the amount of funding that will be necessary for the ongoing management program should be clearly articulated in the bank agreement. If the incentive/outcome-based system is used, the funding to maintain the increased values on the site, on which an increase in credits is based, must also be assured.

The bank agreement should discuss the funding assurances for activities, including habitat management, taking place before, during and after the sale of credits. A management plan should be prepared to help determine the appropriate amount of funding. The management plan should include the activities necessary to implement the biological goals and objectives. Funding for the start-up of the management program should be separate from the requisite endowment for ongoing actions. These initial costs may include upfront costs to the bank owner, including, but not limited to, purchase of the habitat, any enhancements or clean-up required and property taxes. Additionally, there may be consultant or legal fees associated with developing and managing the conservation bank.

Since the management of the bank will be in perpetuity, a good strategy for long-term funding is to establish a non-wasting management endowment (i.e. a fund that generates enough interest each year to cover the costs of the yearly management). This endowment could be established by including the cost of management into the price per credit. As credits are sold, an agreed-upon portion of the proceeds can be deposited into a non-wasting endowment fund or escrow. The size of the required endowment will depend on certain factors that could include the amount of habitat associated with each credit, the land management activities, the amount or degree of habitat restoration needed, the 'risk' of such restoration failing over time, the rate of inflation and the interest rate. For example, low interest rates and a significant active management of the bank lands will require a larger endowment. As a contingency, a time limit should be established for full funding of the non-wasting endowment. The bank owner may have to supplement the endowment at the end of the time limit, if all of the credits have not been sold.

It may also be possible for the conservation bank to support certain agreed-upon revenue generating activities (e.g. bird watching, hiking, grazing, etc.), if these activities do not conflict with the conservation goals of the bank or the intent of the compensation for impacts (e.g. in certain ecological situations, grazing may be a needed management tool). Such monies may be held in escrow or other long-term money management accounts to insure they are available when needed.

Establishment of the conservation bank

A CBA is a legal agreement between the conservation bank owner and a regulatory agency such as the Service or other participating state and/or federal agency that identifies the conditions and criteria under which the bank will be established and operated. The agreement contains information on the exact legal location of the bank and its service area, how credits will be established and managed, and how the bank will be funded, managed and protected in perpetuity. It will deal with issues such as allowable activities and access, and it will identify requirements such as environmental contaminants surveys and appropriate monitoring programs. The CBA itself, once completed, should be signed by the Regional Director.

Management plan

CBAs must include a management plan identifying any habitat or other management activities that will be needed, the endowment necessary to carry out such management in perpetuity, activities allowed to occur on the lands, and monitoring and reporting requirements for management objectives. The bank manager is responsible for fulfilling the obligations of the final management plan. Therefore, it is important to accurately estimate budget needs upfront. If an increase in credits through management actions has been given, the management plan should be updated to reflect the new management needs on the bank. The conservation bank management plan should at a minimum discuss the following issues:

- Property description, including geographical setting, adjacent land uses, location relative to regional open space plans, geology and cultural or historic features on-site.
- Description of biological resources on-site, including vegetation map.
- Identification of activities allowed and prohibited on the conservation bank's land.
- Identification of biological goals and objectives for the bank.

  Management needs of the property, including control of public access, restoration or enhancement of habitats, monitoring of resources, maintenance of facilities, public uses, start-up funding necessary, budget needs and necessary endowment funds to sustain the budget, and yearly reporting requirements. Any special management requirements that are necessary to implement the biological goals and objectives of the bank should also be discussed in detail.
- Any monitoring schedules and special management plan activities, including adaptive management practices.
- Any decision trees or other structures for future management.

Agreement

The main components of a bank agreement are listed below. Because each conservation bank is unique, additional items not listed here may be requested for inclusion in the bank agreement by one or more of the parties as needed. When defining the terms of the bank agreement, keep in mind that both parties' implementation and involvement in the conservation bank will be governed by these terms, unless the conservation bank is further amended by agreement of both parties.

- A general location map and legal description of the property, including GPS coordinates if possible.
- Accurate map(s) of the bank property on a minimum scale of 7 minutes US Geological Survey quad map or finer scale, if available.
- Name of the conservation bank.
- Name of the person(s)/organization(s) to hold fee title to the conservation bank.
- Name of the person(s)/organization(s) with management responsibility for the conservation bank, and for how long. This entity must have demonstrated experience in natural lands management.
- Name of the person or entity that will hold a conservation easement on the property.
- Preliminary title report indicating any easements or encumbrances on the property, including Native American hunting, fishing and gathering rights. This information should be supplied early in the bank evaluation and development process to ensure that the conservation bank's goals are compatible with other current or planned activities on the property.
- An enumeration of the types of potential activities that may include public access and that are compatible with the property's primary function as habitat for species.

- A description of the biological value of the bank, including habitats and species. This may include a vegetation map and biological resources inventory.
- Number and kind of conservation credits within the bank. Final credit numbers and any constraints on types of credits to be sold will be determined by the Service in accordance with a methodology clearly set forth in the agreement.
- An accounting system to track credits, funding and other reporting requirements.
- Description of the service area of the bank. The appropriate service area will be determined by the Service and with the bank owner/manager.
- Description and delineation of each bank phase, if more than one phase is proposed. The description will include phase boundaries, the number of conservation credits associated with each phase, explanation for why the use of phases is preferred, and the agreed upon process for terminating the bank prior to the implementation of all phases.
- Compliance with applicable state and federal laws such as state endangered species acts.
- Results of a Phase I hazardous materials survey for the property.
- A review of mineral and water rights associated with the property.
- Discussion of any prescriptive rights on the property (e.g. road access, etc.).
- An agreement to accurately delineate in the field all boundaries of the bank property, including any bank phases, and construct any required fences before the first conservation credit is sold, fee title transferred or conservation easement granted.
- An agreement to remove any trash, structures or other items on-site that would otherwise reduce the long-term biological value of the site before the first conservation credit is sold, unless otherwise agreed to.
- Provisions for the Service to enter the property for inspections, quality control/assurances and other duties as needed.
- Performance standards that must be achieved.
- Contingency management, funding and ownership plans in the event that the bank owner and/or manager fails to fulfill the obligations as listed under the bank agreement and management plans, including an applicable dispute resolution process to address these contingencies.
- A management plan for the bank property.

Definitions

For the purposes of this guidance document the following terms are defined:

Bank sponsor – any public or private entity responsible for establishing and, in most circumstances, operating a conservation bank.

Conservation actions – the restoration, enhancement or preservation of species habitat for the purpose of reducing adverse impacts to listed species populations.

Conservation bank – a site where habitat and/or other ecosystem resources are conserved and managed in perpetuity for listed species, expressly for the purpose of offsetting impacts occurring elsewhere to the same resource values.

Conservation bank review team – an inter-agency group of federal, state, tribal and/or local regulatory and resource agency representatives who are signatory to a bank agreement and oversee the establishment, use and operation of a conservation bank.

Conservation easement – a recorded legal document established to conserve biological resources in perpetuity, and which requires certain habitat management obligations for the conservation bank lands.

Credit – a unit of measure representing the quantification of species or habitat conservation values within a conservation bank.

Endowment fund – an investment fund maintained by a designated party, approved by the Service, as a non-wasting endowment to be used exclusively for the management of the conservation bank lands in accordance with the management plan and the conservation easement.

Debit – a unit of measure representing the adverse impact to a listed or sensitive species at an impact or project site.

Enhancement – activities conducted in existing species habitat, or other resources, that increase one or more ecosystem functions.

Fee title – a fee title estate is the least limited interest and the most complete and absolute ownership in land; it is of indefinite duration, freely transferable and inheritable.

Management plan – means the plan prepared to manage the conservation bank to, at a minimum, maintain the listed species value on the bank. This includes on-the-ground management activities, funding, and monitoring and reporting requirements.

Non-wasting management endowment – an account that generates enough interest each year to cover the costs of the yearly management.

Off-site conservation – conservation actions occurring outside the boundaries of a project site.

On-site conservation – conservation actions occurring within the boundaries of a project site.

Preservation – the protection of existing ecologically important habitat or other ecosystem resources in perpetuity through the implementation of appropriate legal and physical mechanisms.

Restoration – re-establishment of ecologically important habitat and/or other ecosystem resource characteristics and function(s) at a site where they have ceased to exist, or exist in a substantially degraded state.

Service area – the geographic area (e.g. watershed, county) wherein a bank can reasonably be expected to provide appropriate conservation benefits for impacts to habitat and off-site impacts can be offset by purchase of credits in the bank. The geographic area for which a conservation bank's credits may be applied to offset debits associated with development activities.

Appendix II

Template for a Conservation Bank Agreement

This template provides an example of the content, structure and language currently included in a Conservation Bank Agreement. A standardized template is used to simplify and streamline the bank establishment and management process.

PLEASE NOTE: The following Conservation Bank Agreement is provided as a standardized template document requiring only a limited amount of information to be entered into the spaces where indicated. Under no circumstances will any modifications to this document be accepted beyond the requested information, unless it is addressed within a separately attached amendment hereto, and approved by USFWS.

(Version Date: 10 January 2007)

[Bank Name] CONSERVATION BANK AGREEMENT

This CONSERVATION BANK AGREEMENT ('Agreement') is made and entered into this ____ day of __________ 20__, between ________________________ ________________ ('Owner') and the United States Fish and Wildlife Service (USFWS) (referenced jointly as the 'Parties'). The purpose of this Agreement is to establish the terms and conditions for a Conservation Bank on certain real property to be known as the ___________ Conservation Bank.

I. AGREEMENT RECITALS

WHEREAS:

A. _____________________, [*insert if relevant:* a company incorporated in the State of __________ with an office in __________________,] is the owner ('Owner') of real property located in the County of _______________, including real property located east of ____________________, south of ____________________, north of ____________________, and west of ________________, and more completely described in **Exhibit A** (general and site location maps) and illustrated in **Exhibit B** (title report, legal description and parcel map) attached hereto ('Property').

B. USFWS exercises jurisdiction with respect to the conservation, protection, restoration, enhancement and management of fish, wildlife, native plants and habitat pursuant to various federal laws including the Endangered Species Act, 16 U.S.C. §1531 et seq. ('ESA'), the Fish and Wildlife Coordination Act, 16 U.S.C. §§661–666c, and the Fish and Wildlife Act of 1956, 16 U.S.C. §742(f) et seq.

C. It is anticipated that activities, including construction and development activities ('Activities'), will occur in the 'Service Area' for this Conservation Bank, as set forth and attached hereto as **Exhibit C**, which will necessitate the implementation of mitigation and compensatory conservation measures for impacts to species and habitat subject to this Agreement ('Covered Species').

D. Establishment of this Conservation Bank represents an excellent opportunity to conserve highly valuable biological resources. The Conservation Bank will provide permanent conservation and protection for endangered and threatened species and any additional species the Parties determine will be adequately conserved as a result of implementation of this Agreement, including all life stages and associated habitat, for the endangered: ______________ (*scientific name*), ________________ (*scientific name*); the threatened _____________ (*scientific name*); and the ______________ (*scientific name*) ('Covered Species'). A general description of biological resources, including geographic location and features, topography, vegetation, past and present land use, species, Covered Species, and habitat occurring on the Property, is set forth in the Management Plan at **Exhibit D.** Any recovery plan for any Covered Species shall be referenced in the Management Plan and shall be appended to this Agreement.

E. In accordance with this Agreement, the Parties desire to establish a Conservation Bank on the Property to be known as the '_____________ Conservation Bank' or 'Conservation Bank' to provide for the permanent conservation and management of the Property for the benefit of Covered Species, which shall be used for mitigation and compensatory conservation through the sale or conveyance of 'Conservation Credits' as provided herein.

F. Implementation of this Conservation Bank will be accomplished by recording Conservation Easements in accordance with California Civil Code §815 and this Agreement that will burden the Property in perpetuity to accomplish the conservation and recovery of Covered Species and associated habitat.

G. This Agreement sets forth the terms and conditions by which the Conservation Bank will be established, implemented and governed, as set forth below:

II. AGREEMENT DEFINITIONS

1. The terms used in this Agreement are defined as set forth below:

(1) 'Activity or Activities' means construction, alteration or development including, but not limited to, the construction, development, occupation or operation of industrial, commercial or residential property, as well as any related or associated actions, within the 'Service Area'.

(2) 'Agreement' means this document.

(3) 'Conservation Bank' means the conservation bank established pursuant to this Agreement.

(4) 'Conservation Credit' means a unit of mitigation or compensatory conservation determined by the Parties by use of a proportionate credit formula or other method set forth in the Credit Evaluation attached hereto at **Exhibit E**, which is available to serve as mitigation or compensation for Activities in the Service Area as determined on a project-by-project basis.

(5) 'Conservation Easement' means any conservation easement established in perpetuity and recorded on the Property for conservation purposes consistent with this Agreement and subject to and in accordance with the provisions of California Civil Code §815 and USFWS approval, in the form set forth at **Exhibit F** (including any property description and parcel map), and which shall be attached hereto at **Exhibit F.**

(6) 'Covered Species' means endangered and threatened species and any additional species which the Parties determine will be adequately conserved as a result of implementation of this Agreement, including all life stages and associated habitat, as set forth in the Agreement Recitals and the Management Plan attached hereto at **Exhibit D.**

(7) 'Credit Evaluation' or 'proportionate credit formula' means an evaluation undertaken to formulate the value of each Conservation Credit in relation to the acreage of habitat on a Conservation Easement recorded on the Property (where, for instance, a proportionate credit formula means one acre of habitat on a recorded Conservation Easement equals one Conservation Credit), which shall be attached hereto as **Exhibit E.**

(8) 'Credit Sales Approval' means written approval provided by USFWS to a project proponent indicating that conservation credits purchased from the Conservation Bank may be used as mitigation or compensatory conservation for Activities in the Service Area.

(9) 'Credit Receipt' means a receipt including information in the form attached hereto at **Exhibit I**, executed by Owner and issued for each sale or conveyance of a Conservation Credit.

(10) 'Declaration of Trust' means a declaration by the trustee holding the Endowment Fund, in the form set forth at **Exhibit K**, providing that the Endowment Fund will be utilized for conservation purposes consistent with this Agreement, any Conservation Easement, the Management Plan and any Management Agreement.

(11) 'ESA' means the Endangered Species Act, 16 U.S.C. §1531 et seq.,

including all regulations promulgated pursuant to that Act.

(12) 'Endowment Deposit' means a sum that the Parties determine shall be calculated by a method set forth in **Exhibit J** which shall be collected with the sale or conveyance of each Conservation Credit and deposited into the Endowment Fund until full funding in the Target Amount is achieved for the purpose of perpetually endowing the management of the Conservation Bank for conservation purposes consistent with this Agreement, any Conservation Easement and the Management Plan.

(13) 'Endowment Fund' means a dedicated, non-wasting, interest-bearing account to be established by Owner concurrent with execution of this Agreement and funded with Endowment Deposits collected from the sale or conveyance of Conservation Credits, and which shall generate interest to be used exclusively for management of the Conservation Bank for conservation purposes consistent with this Agreement, any Conservation Easement and the Management Plan.

(14) 'Interim Management Account' means a dedicated interest-bearing account to be established by Owner concurrent with execution of this Agreement and funded with fees collected from the initial sale or conveyance of Conservation Credits in an amount determined by the Parties as set forth at **Exhibit J**, for use as a contingency fund for management of the Conservation Bank until the Endowment Fund Target Amount and Target Date are achieved, at which time the Interim Management Account may be terminated and all funds (excepting interest retained by Owner) shall be transferred to the Endowment Fund for management of the Conservation Bank for one year consistent and in accordance with this Agreement, any Conservation Easement and the Management Plan.

(15) 'Interim Management Period' means the period of time prior to full funding of the Endowment Account in the Target Amount, when Owner shall be responsible for all costs and any management obligations that the Parties agree are to be performed in accordance with this Agreement, any Conservation Easement and the Management Plan.

(16) 'Management Agreement' means an agreement approved by USFWS between the Conservation Bank Owner and an approved party for implementation of the Management Plan utilizing the funding mechanisms established by this Agreement.

(17) 'Management Plan' means the operation and management plan approved by the Parties for the purpose of governing implementation of the terms and conditions of this Agreement, attached hereto as **Exhibit D.**

(18) 'Owner' means the possessor of legal title to the Property subject to this Agreement, its successors or assigns.

(19) 'Property' means Owner's real property upon which Conservation Easements may be recorded and a Conservation Bank may be established as described in this Agreement.

(20) 'Property Assessment and Warranty' means a written assessment of the

number of acres on the Property that Owner warrants may be burdened by Conservation Easement(s) for conservation purposes in accordance with California Civil Code §815 and this Agreement, and which excludes any portion of the Property subject to any right of way, easement, encumbrance or any other matter that prevents establishment of a Conservation Easement on that portion of the Property for conservation purposes in accordance with California Civil Code §815 and this Agreement, and which shall be attached hereto at **Exhibit G**.

(21) 'Service Area' (also known as 'Credit Area') means the geographic area described on the map and in the narrative description set forth at **Exhibit C**, within which impacts that occur may be mitigated or compensated through the use of Conservation Credits.

(22) 'Target Amount' means an amount the Parties determine concurrent with calculation of the Endowment Deposit which is the amount at which the Endowment Fund shall be fully funded, as set forth at **Exhibit J**.

(23) 'Target Date' means the date by which the Parties determine the Endowment Fund shall be fully funded, as set forth at **Exhibit J.**

(24) 'USFWS' means the United States Fish and Wildlife Service, an agency of the United States Department of the Interior.

III. TERMS AND CONDITIONS OF AGREEMENT

NOW, THEREFORE, in consideration of the foregoing Agreement Recitals and other good and valuable consideration, the receipt and sufficiency of which is hereby acknowledged, the Parties agree as follows:

1.0. CONSERVATION BANK EVALUATION AND ACCEPTANCE

1.1. Evaluation of the Property has been undertaken by representatives of USFWS who have inspected and generally evaluated the Property for purposes of determining its biological values in connection with the sale of Conservation Credits. As a result of benefits accruing to wildlife resources including sensitive, threatened and endangered species and their habitats, and particularly to the Covered Species and associated habitat, and upon establishment of the Conservation Bank, USFWS acknowledges that, subject to any limitations provided in this Agreement, the Property possesses biological values sufficient for issuance of Conservation Credits as contemplated by this Agreement.

1.2. Notwithstanding anything to the contrary in this Agreement, in the event that the Property is evaluated and USFWS reasonably determines that portions of the Property have been damaged subsequent to the date of this Agreement and: (1) the effect of such damage has been materially to impair the habitat values on such damaged Property; and (2) Owner has not reasonably restored habitat value to such damaged Property or provided USFWS with reasonable evidence that habitat value will be restored, then USFWS may, at its discretion, either reduce the number of Conservation Credits allocated to the Conservation Bank in proportion to the damaged area or, if USFWS determines that habitat values on the Property have been so impaired as to render the Property unsuitable for use as a Conservation Bank, USFWS may terminate this Agreement.

2.0. ESTABLISHMENT AND DEVELOPMENT OF CONSERVATION BANK

2.1. The Conservation Bank shall be deemed established when:

(a) Owner provides USFWS with general and site location maps **(Exhibit A)**, and a title report, legal description and legal parcel map for the Property **(Exhibit B)**, describing Owner's interest in the Property and documenting any encumbrances (other than monetary encumbrances, which must be subordinated or removed), easements, restrictions or other matters affecting the property.

(b) Owner records a Conservation Easement on the Property in the County where the Property is located which covers the Property or a portion of the Property that USFWS determines constitutes a biologically sustainable unit for conservation purposes contemplated by this Agreement, which shall be in a form subject to USFWS approval and attached hereto as **Exhibit F**, and which shall be held by a party approved by USFWS.

(c) Owner provides USFWS with a 'Property Assessment and Warranty' that identifies the number of acres on the Property that Owner warrants may be burdened by Conservation Easement(s) for conservation purposes in accordance with this Agreement, to be attached hereto as **Exhibit G**.

(d) Owner provides USFWS with a Level I Environmental Contaminants Survey for the Property subject to USFWS approval, to be attached hereto as **Exhibit H**.

(e) The Parties prepare a Management Plan in accordance with this Agreement and Owner satisfies any criteria (such as restoration or enhancement of the Property) required to be performed or completed before Conservation Credits may be sold or conveyed pursuant to the Management Plan.

(f) Owner establishes and deposits funds into an Interim Management Account in an amount the Parties agree is sufficient to fund the Conservation Bank and Management Plan for one year after full funding of the Endowment Fund at the Target Amount has been achieved.

(g) Owner establishes an Endowment Fund account for Endowment Deposits held in trust by an approved party in accordance with the terms and conditions of this Agreement, the Management Plan and any Conservation Easement, subject to approval of USFWS.

2.2. Owner shall provide USFWS with copies of all Conservation Easements recorded on the Property in the form attached hereto as **Exhibit F**, along with a legal parcel map and description of the Property to be covered by any Conservation Easement. Any Conservation Easement shall be attached hereto and incorporated herein as **Exhibit F**, subject to USFWS approval, including approval of the party holding and maintaining the Conservation Easement. When any Conservation Easement is recorded on the Property, Owner shall provide USFWS with an updated Property Assessment of the acres potentially and actually available in the Conservation Bank for conservation purposes in accordance with this Agreement.

2.3. The Property subject to this Agreement may be expanded in accordance with and by amendment to this Agreement upon approval of the Parties, with any expansion of the Property described in a general and site location map, title report, legal description and legal parcel map to be incorporated in **Exhibits A and B** of this Agreement.

2.4. Owner shall pay any and all taxes and charges associated with the Property, obtaining a valid title report, and establishing, recording, selling and conveying any Conservation Credit, Easement or the Conservation Bank.

3.0. CONSERVATION CREDITS

3.1. Upon establishment of the Conservation Bank in accordance with Section 2.0 of this Agreement, Owner may sell and convey Conservation Credits to any project proponent that furnishes a Credit Sales Approval authorizing use of Conservation Credits from the Conservation Bank as mitigation or compensatory conservation for Activities within the Service Area on a project-by-project basis consistent with this Agreement, and Owner shall have the exclusive right to determine the price for any and all Conservation Credits.

3.2. The value of each Conservation Credit in relation to the acreage on a Conservation Easement recorded on the Property shall be determined in accordance with the Credit Evaluation or proportionate credit formula (where, for instance, one acre of habitat on a recorded Conservation Easement equals one Conservation Credit), attached hereto as **Exhibit E**.

3.3. Owner shall issue a 'Credit Receipt' (including information set forth in forms attached hereto at **Exhibit I**), executed by Owner to each purchaser and shall provide a copy of the Credit Receipt to USFWS within thirty (30) days of issuance. If Owner conveys any Conservation Credit to mitigate Activities undertaken on its own property, Owner shall produce a Credit Receipt for the conveyance and shall provide a copy of the Credit Receipt to USFWS.

3.4. Funds from the initial sale or conveyance of Conservation Credits shall be deposited into the Interim Management Account in an amount determined by the Parties in accordance with Section 4.0 of this Agreement, and Credit Receipts may be used as evidence that Conservation Credits in the Conservation Bank have been obtained in accordance with a Credit Sales Approval to satisfy requirements for mitigation or compensatory conservation for Activities in the Service Area.

3.5. Upon each sale or conveyance of a Conservation Credit by Owner, the requisite Endowment Deposit shall be deposited into the Endowment Fund in accordance with Section 4.0 of this Agreement and Owner shall provide USFWS with a copy of the receipt for such Endowment Deposit as provided in Section 4.0 of this Agreement.

3.6. Upon approval of the Parties, the total number of available Conservation Credits may be increased due to a change in methodology used by USFWS to determine habitat value, by the discovery of new biological findings that support an increased number of credits under the current methodology used by USFWS, or by expansion or restoration of the Property, in accordance with and by amendment to this Agreement.

4.0. ENDOWMENT DEPOSITS AND ENDOWMENT FUND

4.1. As a condition of establishment of this Conservation Bank, Owner shall establish a dedicated, non-wasting, interest-bearing account to be funded with Endowment Deposits collected from the sale or conveyance of Conservation Credits, which shall constitute the 'Endowment Fund'. The Endowment Fund shall be held in trust, and Owner shall ensure that any trustee or manager of the Property or Conservation Bank shall utilize the Endowment Fund consistent and in accordance with this Agreement, the Management Plan and any Conservation Easement. Any Declaration of Trust shall be in the form set forth in **Exhibit K,** and is subject to the approval of USFWS.

4.2. Concurrent with the sale or conveyance of each Conservation Credit, Owner shall deposit into the Endowment Fund an 'Endowment Deposit', an amount based upon projected Property management costs (including costs for unforeseen or changed circumstances such as fire, flood, listing of new species) which the Parties calculate by employment of a mutually agreeable method (such as a Property Analysis Record), attached hereto at **Exhibit J.** The calculation of the Endowment Deposit shall account for inflation, which may be incorporated in calculations of projected Property management costs, but may otherwise be revised annually on the execution date of this Agreement beginning in the calendar year following execution of this Agreement in accordance with any increase in the Consumer Price Index (CPI) published by the US Department of Labor, Bureau of Statistics, for the nearest metropolitan area for that year (the CPI adjustment).

4.3. Owner shall provide USFWS with the receipt or other evidence of each Endowment Deposit, which shall be delivered to USFWS with the Credit

Receipt as set forth in Section 3.0, within thirty (30) days of each sale or conveyance of a Conservation Credit. Owner or the Endowment Fund trustee may employ the services of an escrow company to manage the sale of Conservation Credits or Endowment Deposits subject to and in accordance with the terms and conditions of this Agreement.

4.4. In calculating the Endowment Deposit, the Parties shall establish a 'Target Date' for full funding of the Endowment Fund at a 'Target Amount', which shall be determined concurrent with calculation of the Endowment Deposit and shall be set forth and attached hereto at **Exhibit J.** In the event that the Target Amount is not achieved by the Target Date, within thirty (30) days thereafter, Owner shall deposit into the Endowment Fund the difference between the amount in the Endowment Fund and the Target Amount. If the Endowment Deposit has not been calculated to account for inflation **(Exhibit J)**, the Target Amount shall be revised to reflect the result of CPI adjustments, as set forth in this Section. Owner shall ensure that the Target Amount is available upon transfer, assignment or termination of this Agreement. The Target Amount may be revised if the total number of Conservation Credits in the Conservation Bank is increased in accordance with this Agreement.

4.5. As a condition of establishment of this Conservation Bank and before full funding of the Endowment Fund at the Target Amount by the Target Date, Owner shall deposit fees other than Endowment Deposits that are collected from the initial sale or conveyance of Conservation Credits into a dedicated, interest-bearing 'Interim Management Account', in an amount determined by the Parties as set forth at **Exhibit J**, as a contingency fund that may not be expended except by agreement of the Parties and must be replenished if expended. Upon achievement of the Target Amount and Target Date, the Interim Management Account may be terminated and all funds (excepting interest retained by Owner) shall be transferred to the Endowment Fund for management of the Conservation Bank for one year consistent and in accordance with this Agreement, any Conservation Easement and the Management Plan.

4.6. Prior to full funding of the Endowment Account in the Target Amount, Owner shall be responsible for all costs and any management obligations that the Parties agree are to be performed during this 'Interim Management Period' in accordance with this Agreement, any Conservation Easement and the Management Plan. During the Interim Management Period, if the Parties agree that the Interim Management Account may be expended, the Interim Management Account shall be replenished from the succeeding sale of Conservation Credits or by Owner before it is transferred to the Endowment Fund for management of the Conservation Bank. The Parties may, but need not, agree to extend the Target Date in order that sales of Conservation Credits may be used to replenish the Interim Management Account, but Owner shall continue to be responsible for costs and management obligations during this extended Interim Management Period.

4.7. Upon full funding of the Endowment Fund at the Target Amount by the Target Date, the Endowment Fund shall be held and managed for conservation purposes consistent and in accordance with this Agreement, any Conservation Easement and the Management Plan. The accrued interest and earnings from the Endowment Fund shall be used exclusively to fund and defray costs and expenses reasonably incurred for the management of the Property and Conservation Bank, including labor costs, contracts, equipment, materials and signage. Upon satisfaction of the Target Amount, all funds collected from the sale or conveyance of Conservation Credits may be retained by Owner.

4.8. Funds from the Endowment Fund other than interest and earnings may not be expended unless a written request for approval is submitted to USFWS detailing the reasons for the request, and upon written approval by the USFWS, which shall be based upon whether the request is in furtherance of a conservation purpose contemplated by this Agreement and may include provisions for restoration of amounts expended within a prescribed time period.

4.9. Upon transfer, assignment or termination of this Agreement, any funds in the Interim Management Account and the Endowment Fund shall continue to be expended in a manner consistent and in accordance with the conservation purposes for which they were established pursuant to this Agreement and California Civil Code §815, any Conservation Easement and the Management Plan.

5.0. CONSERVATION BANK DATABASE

5.1. A Conservation Bank database, or ledger, shall be established and maintained by Owner for the purpose of tracking funds from the sale and conveyance of Conservation Credits. The database or ledger shall include a numerical accounting of all Conservation Credits available, sold or conveyed, the balance of Conservation Credits remaining, and the aggregate of funds collected, deposited and accrued in any Interim Management Account and the Endowment Fund, as follows:

(a) For each individual sale or conveyance of any Conservation Credit, the database shall state the number and type of Conservation Credits sold or conveyed, the name, address, County and telephone number of the entity purchasing or receiving any Conservation Credit, and the Activity (and the project name and USFWS file number for the project, if available) for which any Conservation Credit was sold or conveyed.

(b) *Within thirty (30) days of each sale or conveyance* of any Conservation Credit, Owner shall provide USFWS with the Credit Receipt and an Endowment Deposit receipt, along with an updated accounting of all funds in any Interim Management Account and all Endowment Deposits collected and deposited into the Endowment Fund, as well as an updated accounting of all Conservation Credits sold or conveyed as of the date of sale or conveyance.

(c) Owner shall provide the database or ledger to USFWS upon request and on an annual basis no later than 15 February, beginning in the calendar year following execution of this Agreement.

(d) Owner shall be responsible for satisfying all of the requirements set forth in this Section until such time as there are no available Conservation Credits remaining for sale or conveyance in the Conservation Bank.

6.0. MANAGEMENT OF CONSERVATION BANK

6.1. Owner, its successors or assigns, shall ensure that the Conservation Bank is managed and maintained consistent and in accordance with this Agreement, any Conservation Easement and the Management Plan.

6.2. As a condition of establishment of this Conservation Bank, Owner and USFWS shall agree upon a Management Plan for the Property, which shall be prepared by Owner and/or an approved manager in consultation with USFWS and shall be attached hereto as **Exhibit D**. The Management Plan shall describe biological resources occurring on the Property and shall provide for the implementation and prioritization of specific management measures and tasks for conservation purposes contemplated by this Agreement, including:

(a) a general description of biological resources, including geographic location and features, topography, vegetation, past and present land use, species, Covered Species and habitat occurring on the Property;

(b) specific measures for ongoing management of the Property and biological resources, as well as measures for Property management should unforeseen or changed circumstances occur, such as fire, flood or listing of new species;

(c) specific measures for regular and ongoing monitoring of the Property to be conducted by a third party monitor expressly approved by USFWS.

6.3. Upon the request of any Party to this Agreement or any Management Agreement, the Parties shall meet and confer to revise the Management Plan to better preserve the habitat and conservation values of the Property. Any changes to the Management Plan shall be subject to USFWS approval and shall be appended to the Management Plan attached hereto as **Exhibit D.**

6.4. Owner and any manager shall ensure that reasonable efforts are employed to prevent third party use of the Property in a manner not in accordance with this Agreement, any Conservation Easement and the Management Plan, including restriction of public access to the Property. Representatives of USFWS shall have a right to enter the Property at any time and USFWS guests may enter with twenty-four (24) hours prior notice by USFWS to Owner or any Management Agreement manager.

6.5. Owner and any Management Agreement manager shall ensure that an accounting and management report is provided to USFWS upon request and on an annual basis no later than 15 February, beginning in the calendar year following execution of this Agreement, which shall include the following:

(a) An accounting of all funds expended in the management of the Property during the previous year;
(b) A general description of the status of the Property;
(c) The results of any biological monitoring or studies conducted on the Property;
(d) A description of all management actions taken on the Property along with a description of any problems encountered in managing the Property; and
(e) A description of management actions that will be undertaken in accordance with this Agreement, any Conservation Easement and the Management Plan in the coming year.

6.6. Upon transfer or assignment of this Agreement or any Management Agreement, all books and records and all rights and responsibilities contained herein shall be transferred or assigned with this Agreement or any Management Agreement and shall continue to be administered consistent and in accordance with this Agreement, any Conservation Easement and the Management Plan.

7.0. OWNER COVENANTS

7.1. Owner hereby agrees and covenants for so long as this Agreement is in effect, that:

(a) Owner shall not discharge or release to the Property, or permit others to discharge or release to the Property, any material or substance deemed 'hazardous' or 'toxic' under any federal, state or local environmental law;
(b) Owner shall not create any encumbrance to the title of the Property other than those set forth in **Exhibit B** and accounted for in the Property Assessment **(Exhibit G),** and Owner shall not execute, renew or extend any lien, license or similar interest without the prior written consent of USFWS;
(c) Owner shall not construct any structure or engage in any activity or use of the Property, including mineral exploration, excavating, draining, dredging or other alteration of the Property, that is not consistent with and in accordance with this Agreement, any Conservation Easement and the Management Plan, without the prior written consent of USFWS; and
(d) Owner shall ensure that the Property is maintained to ensure its suitability as Conservation Bank consistent and in accordance with this Agreement, any Conservation Easement and the Management Plan.

8.0. COOPERATION OF USFWS

8.1. USFWS shall reasonably cooperate with Owner and any approved manager in the implementation of this Agreement. Such cooperation shall include:

(a) Acknowledging to prospective Conservation Credit purchasers that Conservation Credits are available as mitigation or compensatory conservation for Activities within the Service Area;

(b) Acknowledging that the Conservation Bank is 'approved' by USFWS and including the Conservation Bank on a list maintained by USFWS of approved conservation banks and making such list available to prospective Conservation Credit purchasers;

(c) Confirming to prospective Conservation Credit purchasers that purchase of the required number of Conservation Credits from the Conservation Bank shall serve as mitigation or compensatory compensation for incidental take caused by Activities in the Service Area.

9.0. TRANSFER, ASSIGNMENT, TERMINATION OF AGREEMENT

9.1. Owner shall have the right to convey or transfer the Property prior to establishment of the Conservation Bank in accordance with this Agreement and subject to written concurrence by USFWS. If such transfer is made without the prior written concurrence of USFWS, such transfer shall result in the termination of this Agreement.

9.2. Upon establishment of the Conservation Bank pursuant to this Agreement, any transfer or assignment of any portion of, or interest in, the Conservation Bank shall be made only with the prior written concurrence of USFWS, which concurrence shall be subject to the requirement that the successor or assign assume all obligations pursuant to this Agreement and have sufficient financial capacity to carry out any unfunded Agreement obligations. Transfer or assignment of this Agreement to a party approved in writing by USFWS shall also be subject to the requirement that any funds in an Interim Management Account and the Endowment Fund shall continue to be expended in a manner consistent and in accordance with this Agreement, any Conservation Easement and the Management Plan.

9.3. USFWS may terminate this Agreement on the condition that each of the following has occurred: (i) Owner has breached one or more Owner Covenants or terms and conditions set forth herein; (ii) Owner has received written notice of such breach from USFWS; and (iii) Owner has failed to cure such breach within thirty (30) days after such notice; provided that in the event such breach is curable in the judgment of USFWS, but cannot reasonably be cured within such thirty (30) day period, USFWS shall not terminate this Agreement so long as Owner has commenced the cure of such breach and is diligently pursuing such cure to completion. Nothing in this paragraph is intended or shall be construed to limit the legal or equitable remedies (including specific performance and injunctive relief) at law available to USFWS in the event of a threatened or actual breach of this Agreement. If this Agreement is terminated, funds in an Interim Management Account and the Endowment Fund shall continue to be expended in a manner consistent and in accordance with the conservation purposes for which they were established pursuant to this Agreement and California Civil Code §815, any Conservation Easement and the Management Plan, and any Conservation Easement shall continue in perpetuity as a covenant running with the land.

9.4. Upon written concurrence of USFWS, this Agreement may be terminated following the sale or conveyance of all available Conservation Credits and satisfaction of all substantive terms and conditions of this Agreement other than ongoing management obligations such as maintenance and monitoring, as set forth in the Management Plan, provided that the Endowment Fund shall continue to be expended in a manner consistent and in accordance with the conservation purposes for which it was established pursuant to this Agreement and California Civil Code §815, any Conservation Easement and the Management Plan, and any Conservation Easement shall continue in perpetuity as a covenant running with the land.

10.0. REMEDIES AND ENFORCEMENT OF AGREEMENT

10.1. The Parties shall each have all of the remedies available in equity (including specific performance and injunctive relief) and at law to enforce the terms of this Agreement and to seek remedies for any breach or violation thereof. Nothing in this Agreement shall be deemed to limit USFWS jurisdiction over endangered, threatened and sensitive species and biological resources, or to restrict the ability of USFWS to seek civil or criminal penalties or otherwise fully discharge its responsibilities under applicable law including the ESA.

10.2. The Parties agree to work together in good faith to resolve disputes concerning this Agreement but any Party may seek any available remedy. Unless an aggrieved party has initiated administrative proceedings or suit in federal court, the Parties may elect to employ an informal dispute resolution process whereby:

(a) The aggrieved Party shall notify any other Party of the Provision that may have been violated, the basis for contending that a violation has occurred, and the remedies it proposes to correct the alleged violation;

(b) The Party alleged to be in violation shall have thirty (30) days or such other time as may be agreed upon to respond and, during this time, may seek clarification of the initial notice and shall use its best efforts to provide any responsive information;

(c) Within thirty (30) days after such response was provided or due, Party representatives shall confer and negotiate in good faith toward a resolution satisfactory to each Party, or shall establish a specific process and timetable to seek such solution.

11.0. ENTIRE AGREEMENT

11.1. This Agreement constitutes the entire agreement of the Parties, and no other agreement, statement or promise made by the Parties, or to any employee, officer or agent of the Parties, which is not contained in this Agreement shall be binding or valid.

11.2. The template form of this Agreement is not subject to amendment or modification except by written consent of the Parties and any attempted modification not in compliance with this requirement shall be void. Any amendment or modification to this template Agreement shall be included in an addendum or an Exhibit to this Agreement with reference to the specific provisions modified or deleted.

11.3. All Exhibits referred to in this Agreement are attached to this Agreement and are incorporated herein by reference.

12.0. SUCCESSORS AND ASSIGNS

12.1. This Agreement and each of its covenants and conditions shall be binding on and shall inure to the benefit of the Parties and their respective successors and assigns.

12.2. Owner may transfer or assign its rights and obligations under this Agreement consistent with applicable USFWS regulations, this Agreement, and with the prior written approval of USFWS, which approval shall not be unreasonably withheld.

13.0. NOTICE

13.1. Any notice, demand or request permitted or required by this Agreement shall be delivered personally, sent by facsimile, or sent by recognized overnight delivery service, to the persons in the positions set forth below or shall be deemed given five (5) days after deposit in the United States mail, certified and postage prepaid, return receipt requested, and addressed as follows or at such other address as either the Parties may from time to time specify in writing:

Owner(s): ______________________________

____

__________________, CA _____

Attention: ______________________________;

USFWS: United States Fish and Wildlife Service
Sacramento Fish and Wildlife Office

2800 Cottage Way, Room W-2605
Sacramento, CA 95825-1846
Attention: Field Supervisor

14.0. ATTORNEY FEES

14.1. If any action at law or equity, including any action for declaratory relief, is brought to enforce or interpret the provisions of this Agreement, each Party to the litigation shall bear its own attorney fees and costs.

15.0. AVAILABILITY OF FUNDS

15.1. Implementation of this Agreement by USFWS is subject to the requirements of the Anti-Deficiency Act, 31 U.S.C. §1341, and the availability of appropriated funds. Nothing in this Agreement may be construed to require the obligation, appropriation or expenditure of any money from the United States Treasury. USFWS is not required under this Agreement to expend any appropriated funds unless and until an authorized official affirmatively acts to commit to such expenditures as evidenced in writing.

16.0. ELECTED OFFICIALS

16.1. No member or delegate to Congress shall be entitled to any share or part of this Agreement, or to any benefit that may arise from it.

17.0. NO PARTNERSHIPS

17.1. This Agreement shall not make or be deemed to make any Party to this Agreement an agent for or the partner of any other Party.

18.0. GOVERNING LAW

18.1. This Agreement shall be governed by and construed in accordance with the Federal Endangered Species Act, 16 U.S.C. §§661–666c, the Fish and Wildlife Act of 1956, 16 U.S.C. §742(f) et seq., the laws of the State of California, and other applicable federal laws and regulations. Nothing in this Agreement is intended to limit the authority of USFWS to seek penalties or otherwise fulfill its responsibilities under the ESA or as an agency of the Federal government.

19.0. COUNTERPARTS

19.1. This Agreement may be executed in identical counterparts and each counterpart shall be deemed to be an original document. All executed counterparts together shall constitute one and the same document, and any counterpart signature pages may be detached and assembled to form a single original document.

AGREEMENT EXHIBITS

The following Exhibits are attachments incorporated in this Conservation Bank Agreement:

Exhibit A	General Location Map and Site Location Map for Property
Exhibit B	Preliminary Title Report, Legal Description and Legal Parcel Map for Property
Exhibit C	Service Area (or 'Credit Area') Map and Text Description
Exhibit D	Management Plan (including Economic Analysis), Biological Surveys, Verified Wetland Delineation (if required)
Exhibit E	Credit Evaluation
Exhibit F	Conservation Easement(s), Parcel Map(s), Property Description(s), Subordination Agreement (when required) and Title Insurance
Exhibit G	Property Assessment and Warranty
Exhibit H	Level I Environmental Assessment
Exhibit I	Credit Receipt Instructions and Forms
Exhibit J	Endowment Deposit, Target Amount, Target Date, Interim Management Account
Exhibit K	Declaration of Trust

IN WITNESS WHEREOF, THE PARTIES HERETO have executed this ______ __ Conservation Bank Agreement as of the date last signed below.

OWNER

By:____
Date:______________________

Title:______________________________________

UNITED STATES FISH AND WILDLIFE SERVICE

By:____
Date:______________________

Title: Field Supervisor, Sacramento Fish and Wildlife Office

Appendix III

Template for a Bank's Conservation Easement

This template provides an example of the content, structure and language currently included in a conservation bank's conservation easement. A standardized template is used to simplify and streamline the bank establishment and management process.

CONSERVATION EASEMENT DEED
________________________ Mitigation Bank/Project
[When DFG is NOT Grantee]

THIS CONSERVATION EASEMENT DEED ('Conservation Easement') is made as of the ______ day of ________________, 20____, by ____________ ______________ ('Grantor'), in favor of ______________________________ ('Grantee'), with reference to the following facts:

RECITALS

A. Grantor is the sole owner in fee simple of certain real property containing approximately ______ acres, located in the [***City of***____________ ____,] County of ______________, State of California , and including designated Assessor's Parcel Number(s) ______________ (the 'Property'). The Property is legally described in **Exhibit A** and depicted on the map in **Exhibit B** attached to this Conservation Easement and incorporated in it by this reference.

B. The Property possesses wildlife and habitat values (collectively, 'Conservation Values') of great importance to Grantee, the people of the State of California and the people of the United States, including, among other things, the specific Conservation Values identified in Recital C, below.

C. The Property provides high quality habitat for [***list plant and/or animal species***] and contains [***list habitats; native and/or non-native***], and restored, created, enhanced and/or preserved jurisdictional waters of the United States including wetlands.

D. Grantee is authorized to hold easements pursuant to CA Civil Code §815.3. Specifically, Grantee is [*choose applicable statement: a tax-exempt nonprofit organization qualified under section 501(c) (3) of the Internal Revenue Code of 1986, as amended, and qualified to do business in California which has as its primary purpose the preservation of land in its natural, scenic, forested or open space condition or use OR a governmental entity identified in CA Civil Code Section 815.3(b) and otherwise authorized to acquire and hold title to real property*].

E. The United States Fish and Wildlife Service ('USFWS'), an agency within the United States Department of the Interior, has jurisdiction over the conservation, protection, restoration and management of fish, wildlife, native plants and the habitat necessary for biologically sustainable populations of these species within the United States pursuant to the Endangered Species Act, 16 U.S.C. §1531, *et seq.*, the Fish and Wildlife Coordination Act, 16 U.S.C. §§661–666c, the Fish and Wildlife Act of 1956, 16 U.S.C. §742(f), *et seq.*, and other provisions of federal law.

F. This Conservation Easement provides mitigation for impacts of approved projects affecting wetlands and associated habitats and species located [***insert general service area description***], County of [***insert name***], State of California, pursuant to the Mitigation Bank Enabling Instrument for the [***insert name***] Mitigation Bank, CDFG Tracking No. [***insert number***] (the '[BEI *or* Conservation Bank Agreement]'), by and between [***insert Owner's name(s)***] and CDFG, the USFWS, USFWS File No. [***insert number***], the ________________ District of the US Army Corps of Engineers ('USACE'), USACE File Number [*insert number*], and Region [*insert number*] of the US Environmental Protection Agency ('USEPA'), entered into concurrently with this Conservation Easement, and the Bank Development Plan (the 'Development Plan') and Bank Management Plan (the 'Management Plan') created under the [BEI *or* Conservation Bank Agreement]. CDFG, USFWS, USACE and USEPA are together referred to in this Conservation Easement as the 'Signatory Agencies'.

COVENANTS, TERMS, CONDITIONS AND RESTRICTIONS

For good and valuable consideration, the receipt and sufficiency of which is hereby acknowledged, and pursuant to the laws of the United States and the State of California, including California Civil Code §815, *et seq.*, Grantor hereby voluntarily grants and conveys to Grantee a conservation easement in perpetuity over the Property.

1. Purposes. The purposes of this Conservation Easement are to ensure that the Property will be retained forever in its natural, restored or enhanced condition as contemplated by the [BEI *or* Conservation Bank Agreement], the Development Plan and the Management Plan, and to prevent any use of the Property that will impair or interfere with the Conservation Values of the Property as so restored or enhanced. Grantor intends that this Conservation Easement will confine the use of the Property to activities that are consistent with such purposes, including, without limitation, those involving the preservation, restoration and enhancement of native species and their habitats implemented in accordance with the [BEI *or* Conservation Bank Agreement], the Development Plan and the Management Plan.

   A final, approved copy of the [BEI *or* Conservation Bank Agreement], the Development Plan and the Management Plan, and any amendments thereto approved by the Signatory Agencies, shall be kept on file at the respective offices of the Signatory Agencies. If Grantor, or any successor or assign, requires an official copy of the [BEI *or* Conservation Bank Agreement], the Development Plan or the Management Plan, it should request a copy from one of the Signatory Agencies at its address for notices listed in Section 12 of this Conservation Easement.

   The [BEI *or* Conservation Bank Agreement], the Development Plan and the Management Plan are incorporated by this reference into this Conservation Easement as if fully set forth herein.

2. Grantee's Rights. To accomplish the purposes of this Conservation Easement, Grantor hereby grants and conveys the following rights to Grantee and to USFWS as a third-party beneficiary of this Conservation Easement:

(a) To preserve and protect the Conservation Values of the Property.

(b) To enter the Property at reasonable times in order to monitor compliance with and otherwise enforce the terms of this Conservation Easement, the [BEI *or* Conservation Bank Agreement], the Development Plan and the Management Plan; and to implement at Grantee's sole discretion Development Plan and Management Plan activities that have not been implemented, provided that Grantee shall not unreasonably interfere with Grantor's authorized use and quiet enjoyment of the Property.

(c) To prevent any activity on or use of the Property that is inconsistent with the purposes of this Conservation Easement and to require the restoration of such areas or features of the Property that may be damaged by any act, failure to act or any use or activity that is inconsistent with the purposes of this Conservation Easement.

(d) The right to require that all mineral, air and water rights as Grantee deems necessary to preserve and protect the biological resources and Conservation Values of the Property shall remain a part of and be put to beneficial use upon the Property, consistent with the purposes of this Conservation Easement.

(e) All present and future development rights appurtenant to, allocated, implied, reserved or inherent in the Property; such rights are hereby terminated and extinguished, and may not be used on or transferred to any portion of the Property, nor any other property adjacent or otherwise.

3. Prohibited Uses. Any activity on or use of the Property that is inconsistent with the purposes of this Conservation Easement is prohibited. Without limiting the generality of the foregoing, the following uses and activities by Grantor, Grantor's agents and third parties are expressly prohibited:

(a) Unseasonable watering; use of fertilizers, pesticides, biocides, herbicides or other agricultural chemicals; weed abatement activities; incompatible fire protection activities; and any and all other activities and uses which may adversely affect the purposes of this Conservation Easement[, ***except as otherwise specifically provided in the Development Plan or the Management Plan***].

(b) Use of off-road vehicles and use of any other motorized vehicles except on existing roadways[, ***except as otherwise specifically provided in the Development Plan or the Management Plan***].

(c) Agricultural activity of any kind[, ***except grazing for vegetation management as specifically provided in the Development Plan or the Management Plan***].

(d) Recreational activities, including, but not limited to, horseback riding, biking, hunting or fishing[, ***except for personal, non-commercial, recreational activities of the Grantor, so long as such activities are consistent with the purposes of this Conservation Easement, and recreational activities (if any) as specifically provided in the Development Plan or the Management Plan***].

(e) Commercial or industrial uses.

(f) Any legal or de facto division, subdivision or partitioning of the Property.

(g) Construction, reconstruction, erecting or placement of any building, billboard or sign, or any other structure or improvement of any kind[, ***except as otherwise specifically provided in the Development Plan or the Management Plan].***

(h) Depositing or accumulation of soil, trash, ashes, refuse, waste, bio-solids or any other materials.

(i) Planting, introduction or dispersal of non-native or exotic plant or animal species.

(j) Filling, dumping, excavating, draining, dredging, mining, drilling, removing or exploring for or extracting minerals, loam, soil, sands, gravel, rocks or other material on or below the surface of the Property, or granting or authorizing surface entry for any of these purposes.

(k) Altering the surface or general topography of the Property, including building roads or trails, paving or otherwise covering the Property with concrete, asphalt or any other impervious material[, ***except as otherwise specifically provided in the Development Plan or the Management Plan***].

(l) Removing, destroying, or cutting of trees, shrubs or other vegetation, except as required by law for (1) fire breaks, (2) maintenance of existing foot trails or roads, (3) prevention or treatment of disease[***, or as otherwise specifically provided in the Development Plan or the Management Plan***].

(m) Manipulating, impounding or altering any natural water course, body of water or water circulation on the Property, and any activities or uses detrimental to water quality, including but not limited to degradation or pollution of any surface or sub-surface waters[***, except as otherwise specifically provided in the Development Plan or the Management Plan***].

(n) Without the prior written consent of Grantee, which Grantee may withhold, transferring, encumbering, selling, leasing or otherwise separating the mineral rights or water rights for the Property; changing the place or purpose of use of the water rights; abandoning or allowing the abandonment of, by action or inaction, any water or water rights, ditch or ditch rights, spring rights, reservoir or storage rights, wells, ground water rights, or other rights in and to the use of water historically used on or otherwise appurtenant to the Property.

(o) Engaging in any use or activity that may violate, or may fail to comply with, any relevant federal, state or local laws, regulations and policies applicable to Grantee, the Property, or the use or activity in question.

4. Grantor's Duties.

(a) Grantor shall undertake all reasonable actions to prevent the unlawful entry and trespass by persons whose activities may degrade or harm the Conservation Values of the Property or that are otherwise inconsistent with this Conservation Easement. In addition, Grantor shall undertake all necessary actions to perfect and defend rights of Grantee and third-party beneficiaries under Section 2 of this Conservation Easement, and to implement the [BEI *or* Conservation Bank Agreement], the Development Plan and the Management Plan.

(b) Grantor shall not transfer, encumber, sell, lease or otherwise separate the mineral, air or water rights for the Property, or change the place or purpose of use of the water rights, without first obtaining the written consent of Grantee, which Grantee may withhold. Grantor shall not abandon or allow the abandonment of, by action or inaction, any of Grantor's right, title or interest in and to any water or water rights, ditch or ditch rights, spring rights, reservoir or storage rights, wells, ground water rights, or other rights in and to the use of water historically used on or otherwise appurtenant to the Property including, without limitation: (i) riparian water rights; (ii) appropriative water rights; (iii) rights to waters which are secured under contract with any irrigation or water district, to the extent such waters are customarily applied to the Property; or (iv) any water from wells that are in existence or may be constructed in the future on the Property.

(c) Grantor shall install and maintain a fence reasonably satisfactory to

Grantee around the Property to protect the Conservation Values of the Property, including but not limited to wildlife corridors.

5. Reserved Rights. Grantor reserves to itself, and to its personal representatives, heirs, successors and assigns, all rights accruing from Grantor's ownership of the Property, including the right to engage in or permit or invite others to engage in all uses of the Property that are not prohibited or limited by, and are consistent with the purposes of, this Conservation Easement.

6. Grantee's Remedies. USFWS, as a third-party beneficiary under this Conservation Easement, shall have the same rights as Grantee under this section to enforce the terms of this Conservation Easement. If Grantee determines that a violation of the terms of this Conservation Easement has occurred or is threatened, Grantee shall give written notice to Grantor of such violation and demand in writing the cure of such violation. If Grantor fails to cure the violation within thirty (30) days after receipt of written notice and demand from Grantee, or if the cure reasonably requires more than thirty (30) days to complete and Grantor fails to begin the cure within the thirty (30)-day period or fails to continue diligently to complete the cure, Grantee may bring an action at law or in equity in a court of competent jurisdiction to enforce this Conservation Easement, to recover any damages to which Grantee may be entitled for violation of the terms of this Conservation Easement or for any injury to the Conservation Values of the Property, to enjoin the violation, *ex parte* as necessary, by temporary or permanent injunction without the necessity of proving either actual damages or the inadequacy of otherwise available legal remedies, or for other equitable relief, including, but not limited to, the restoration of the Property to the condition in which it existed prior to any violation or injury. Without limiting the liability of Grantor, Grantee may apply any damages recovered to the cost of undertaking any corrective action on the Property.

If Grantee, in its sole discretion, determines that circumstances require immediate action to prevent or mitigate damage to the Conservation Values of the Property, Grantee may pursue its remedies under this Conservation Easement without prior notice to Grantor or without waiting for the period provided for cure to expire. Grantee's rights under this section apply equally to actual or threatened violations of the terms of this Conservation Easement. Grantor agrees that Grantee's remedies at law for any violation of the terms of this Conservation Easement are inadequate and that Grantee shall be entitled to the injunctive relief described in this section, both prohibitive and mandatory, in addition to such other relief to which Grantee may be entitled, including specific performance of the terms of this Conservation Easement, without the necessity of proving either actual damages or the inadequacy of otherwise available legal remedies. Grantee's remedies described in this section shall be cumulative and shall be in addition to all remedies now or hereafter existing at law or in equity, including but not limited to, the remedies set forth in Civil Code §815, *et seq*. The failure of Grantee to discover a violation or to take immediate legal action shall not bar Grantee from taking such action at a later time.

If at any time in the future Grantor or any successor in interest or subsequent transferee uses or threatens to use the Property for purposes inconsistent with or in violation of this Conservation Easement then, notwithstanding Civil Code §815.7, CDFG, the California Attorney General or any third-party beneficiary of this Conservation Easement has standing as an interested party in any proceeding affecting this Conservation Easement.

6.1. Costs of Enforcement. All costs incurred by Grantee, where Grantee is the prevailing party, in enforcing the terms of this Conservation Easement against Grantor, including, but not limited to, costs of suit and attorneys' and experts' fees, and any costs of restoration necessitated by negligence or breach of this Conservation Easement shall be borne by Grantor.

6.2. Grantee's Discretion. Enforcement of the terms of this Conservation Easement by Grantee or USFWS shall be at the discretion of Grantee or USFWS, and any forbearance by Grantee or USFWS to exercise its rights under this Conservation Easement in the event of any breach of any term of this Conservation Easement shall not be deemed or construed to be a waiver of such term or of any subsequent breach of the same or any other term of this Conservation Easement or of any rights of Grantee (or any rights of USFWS, as a third-party beneficiary) under this Conservation Easement. No delay or omission by Grantee or USFWS in the exercise of any right or remedy shall impair such right or remedy or be construed as a waiver.

6.3. Acts Beyond Grantor's Control. Nothing contained in this Conservation Easement shall be construed to entitle Grantee to bring any action against Grantor for any injury to or change in the Property resulting from (i) any natural cause beyond Grantor's control, including, without limitation, fire not caused by Grantor, flood, storm, and earth movement, or any prudent action taken by Grantor under emergency conditions to prevent, abate or mitigate significant injury to the Property resulting from such causes; or (ii) acts by Grantee or its employees.

6.4. USFWS Right of Enforcement. All rights and remedies conveyed to Grantee under this Conservation Easement shall extend to and are enforceable by USFWS. These rights are in addition to, and do not limit, the rights of enforcement under the [BEI *or* Conservation Bank Agreement], the Development Plan or the Management Plan.

7. Fence Installation and Maintenance. Grantor agrees to install and maintain a fence as described in section 4, Grantor's Duties.

8. Access. This Conservation Easement does not convey a general right of access to the public.

9. Costs and Liabilities. Grantor retains all responsibilities and shall

bear all costs and liabilities of any kind related to the ownership, operation, upkeep and maintenance of the Property. Grantor agrees that neither Grantee nor USFWS shall have any duty or responsibility for the operation, upkeep or maintenance of the Property, the monitoring of hazardous conditions on it, or the protection of Grantor, the public or any third parties from risks relating to conditions on the Property. Grantor remains solely responsible for obtaining any applicable governmental permits and approvals required for any activity or use permitted by this Conservation Easement, including permits and approvals required from Grantee acting in its regulatory capacity, and any activity or use shall be undertaken in accordance with all applicable federal, state, local and administrative agency laws, statutes, ordinances, rules, regulations, orders and requirements.

9.1. Taxes; No Liens. Grantor shall pay before delinquency all taxes, assessments (general and special), fees and charges of whatever description levied on or assessed against the Property by competent authority (collectively 'Taxes'), including any Taxes imposed upon, or incurred as a result of, this Conservation Easement, and shall furnish Grantee or USFWS with satisfactory evidence of payment upon request. Grantor shall keep the Property free from any liens (other than a security interest that is expressly subordinated to this Conservation Easement, as provided in Section 14 (k)), including those arising out of any obligations incurred by Grantor for any labor or materials furnished or alleged to have been furnished to or for Grantor at or for use on the Property.

9.2. Hold Harmless. Grantor shall hold harmless, protect and indemnify Grantee and its directors, officers, employees, agents, contractors, and representatives and the heirs, personal representatives, successors and assigns of each of them (each an 'Indemnified Party' and, collectively, 'Indemnified Parties') from and against any and all liabilities, penalties, costs, losses, damages, expenses (including, without limitation, reasonable attorneys' fees and experts' fees), causes of action, claims, demands, orders, liens or judgments (each a 'Claim' and, collectively, 'Claims'), arising from or in any way connected with: (a) injury to or the death of any person, or physical damage to any property, resulting from any act, omission, condition or other matter related to or occurring on or about the Property, regardless of cause, unless due solely to the negligence of Grantee or any of its employees; (b) the obligations specified in Sections 4, 9 and 9.1; and (c) the existence or administration of this Conservation Easement. If any action or proceeding is brought against any of the Indemnified Parties by reason of any such Claim, Grantor shall, at the election of and upon written notice from Grantee, defend such action or proceeding by counsel reasonably acceptable to the Indemnified Party or reimburse Grantee for all charges incurred for services of the Attorney General in defending the action or proceeding.

9.3. Extinguishment. If circumstances arise in the future that render the purposes of this Conservation Easement impossible to accomplish, this Conservation Easement can only be terminated or extinguished, in whole or in part, by judicial proceedings in a court of competent jurisdiction.

9.4. Condemnation. The purposes of this Conservation Easement are presumed to be the best and most necessary public use as defined at CA Code of Civil Procedure Section 1240.680 notwithstanding CA Code of Civil Procedure Sections 1240.690 and 1240.700.

10. Transfer of Conservation Easement. This Conservation Easement may be assigned or transferred by Grantee upon written approval of the Signatory Agencies, which approval shall not be unreasonably withheld or delayed, but Grantee shall give Grantor and the Signatory Agencies at least thirty (30) days prior written notice of the transfer. Approval of any assignment or transfer may be withheld in the reasonable discretion of the Signatory Agencies if the transfer will result in a single owner holding both this Conservation Easement and fee title to the Property and, upon such transfer, the doctrine of merger would apply to extinguish the Conservation Easement by operation of law, unless, prior to assignment of transfer, an alternate method or mechanism to achieve the purposes of this Conservation Easement following such merger has been provided for. Grantee may assign or transfer its rights under this Conservation Easement only to an entity or organization authorized to acquire and hold conservation easements pursuant to Civil Code §815.3 (or any successor provision then applicable) or the laws of the United States and reasonably acceptable to the Signatory Agencies. Grantee shall require the assignee to record the assignment in the county where the Property is located. The failure of Grantee to perform any act provided in this section shall not impair the validity of this Conservation Easement or limit its enforcement in any way.

11. Transfer of Property. Grantor agrees to incorporate the terms of this Conservation Easement by reference in any deed or other legal instrument by which Grantor divests itself of any interest in all or any portion of the Property, including, without limitation, a leasehold interest. Grantor agrees that the deed or other legal instrument shall also incorporate by reference the [BEI *or* Conservation Bank Agreement], the Development Plan, the Management Plan and any amendment(s) to those documents. Grantor further agrees to give written notice to Grantee and the Signatory Agencies of the intent to transfer any interest at least thirty (30) days prior to the date of such transfer. Grantee or the Signatory Agencies shall have the right to prevent subsequent transfers in which prospective subsequent claimants or transferees are not given notice of the terms, covenants, conditions and restrictions of this Conservation Easement (including the exhibits and documents incorporated by reference in it). If Grantor proposes to transfer fee title to the Property to the then Grantee of this Conservation Easement, and if the doctrine of merger would apply and extinguish the Conservation Easement by operation of law upon such transfer, then the transfer shall be subject to

the prior written approval of the Signatory Agencies, which approval shall not be unreasonably withheld or delayed. Approval of any such transfer that is subject to the approval of the Signatory Agencies may be withheld in the reasonable discretion of the Signatory Agencies unless, prior to such transfer, an alternate method or mechanism to achieve the purposes of this Conservation Easement following such merger has been provided for. The failure of Grantor to perform any act provided in this section shall not impair the validity of this Conservation Easement or limit its enforceability in any way.

12. Notices. Any notice, demand, request, consent, approval or other communication that Grantor or Grantee desires or is required to give to the other shall be in writing, with a copy to each of the Signatory Agencies, and be served personally or sent by recognized overnight courier that guarantees next-day delivery or by first class United States mail, postage fully prepaid, addressed as follows:

To Grantor: ______________________________
____
____
____

To Grantee: ______________________________
____
____
____

To CDFG: Department of Fish and Game
______________________ Region
[*Insert address*]
________________, CA ________
Attn: Regional Manager

With a copy to: Department of Fish and Game
Office of General Counsel
1416 Ninth Street, 12th Floor
Sacramento, CA 95814-2090
Attn: General Counsel

To USFWS: United States Fish and Wildlife Service
2800 Cottage Way, W-2605
Sacramento, CA 95826-1846
Attn: Field Supervisor

To USACE: US Army Corps of Engineers
____________________ District
Attn: ____________________
____

To USEPA: US Environmental Protection Agency, Region 9
Attn: Director, Water Division
75 Hawthorne Street
San Francisco, CA 94105

or to such other address a party or a Signatory Agency shall designate by written notice to Grantor, Grantee and the Signatory Agencies. Notice shall be deemed effective upon delivery in the case of personal delivery or delivery by overnight courier or, in the case of delivery by first class mail, five (5) days after deposit into the United States mail.

13. Amendment. This Conservation Easement may be amended only by mutual written agreement of Grantor and Grantee, and written approval of the Signatory Agencies (which approval shall not be unreasonably withheld or delayed). Any such amendment shall be consistent with the purposes of this Conservation Easement and California law governing conservation easements and shall not affect its perpetual duration. Any such amendment shall be recorded in the official records of the county in which the Property is located, and Grantee shall promptly provide a conformed copy of the recorded amendment to the Grantor and the Signatory Agencies.

14. Additional Provisions.

(a) Controlling Law. The interpretation and performance of this Conservation Easement shall be governed by the laws of the State of California, disregarding the conflicts of law principles of such state, and applicable federal law, including the ESA.

(b) Liberal Construction. Despite any general rule of construction to the contrary, this Conservation Easement shall be liberally construed to effect the purposes of this Conservation Easement and the policy and purpose of Civil Code §815, *et seq*. If any provision in this instrument is found to be ambiguous, an interpretation consistent with the purposes of this Conservation Easement that would render the provision valid shall be favored over any interpretation that would render it invalid.

(c) Severability. If a court of competent jurisdiction voids or invalidates on its face any provision of this Conservation Easement, such action shall not affect the remainder of this Conservation Easement. If a court of competent jurisdiction voids or invalidates the application of any provision of this Conservation Easement to a person or circumstance, such action shall not affect the application of the provision to any other persons or circumstances.

(d) Entire Agreement. This instrument (including its exhibits and any [BEI *or* Conservation Bank Agreement], Development Plan, Management Plan, and endowment fund incorporated by reference in it) sets forth the entire agreement of the parties and the Signatory Agencies with respect to the Conservation Easement and supersedes all prior discussions, negotiations, understandings or agreements relating to the Conservation Easement. No alteration or variation of this instrument shall be valid or binding unless contained in an amendment in accordance with Section 13.

(e) No Forfeiture. Nothing contained herein will result in a forfeiture or reversion of Grantor's title in any respect.

(f) Successors. The covenants, terms, conditions and restrictions of this Conservation Easement shall be binding upon, and inure to the benefit of, the parties hereto and their respective personal representatives, heirs, successors and assigns and shall constitute a servitude running in perpetuity with the Property.

(g) Termination of Rights and Obligations. A party's rights and obligations under this Conservation Easement terminate upon transfer of the party's interest in the Conservation Easement or Property, except that liability for acts, omissions or breaches occurring prior to transfer shall survive transfer.

(h) Captions. The captions in this instrument have been inserted solely for convenience of reference and are not a part of this instrument and shall have no effect upon its construction or interpretation.

(i) No Hazardous Materials Liability. Grantor represents and warrants that it has no knowledge or notice of any Hazardous Materials (defined below) or underground storage tanks existing, generated, treated, stored, used, released, disposed of, deposited or abandoned in, on, under or from the Property, or transported to or from or affecting the Property. Without limiting the obligations of Grantor under Section 9.2, Grantor hereby releases and agrees to indemnify, protect and hold harmless the Indemnified Parties (defined in Section 9.2) from and against any and all Claims (defined in Section 9.2) arising from or connected with any Hazardous Materials or underground storage tanks present, alleged to be present, released in, from or about, or otherwise associated with the Property at any time, except any Hazardous Materials placed, disposed or released by Grantee, its employees or agents. This release and indemnification includes, without limitation, Claims for injury to or death of any person or physical damage to any property; and the violation or alleged violation of, or other failure to comply with, any Environmental Laws (defined below). If any action or proceeding is brought against any of the Indemnified Parties by reason of any such Claim, Grantor shall, at the election of and upon written notice from Grantee, defend such action or proceeding by counsel reasonably acceptable to the Indemnified Party or reimburse Grantee for all charges incurred for services of the Attorney General in defending the action or proceeding.

Despite any contrary provision of this Conservation Easement, the parties do not intend this Conservation Easement to be, and this Conservation Easement shall not be, construed such that it creates in or gives to Grantee any of the following:

(1) The obligations or liability of an 'owner' or 'operator', as those terms are defined and used in Environmental Laws (defined below), including, without limitation, the Comprehensive Environmental Response, Compensation and Liability Act of 1980, as amended (42 U.S.C. §9601, *et seq*.; hereinafter, 'CERCLA'); or
(2) The obligations or liabilities of a person described in 42 U.S.C. §9607(a) (3) or (4); or
(3) The obligations of a responsible person under any applicable Environmental Laws; or
(4) The right to investigate and remediate any Hazardous Materials associated with the Property; or
(5) Any control over Grantor's ability to investigate, remove, remediate or otherwise clean up any Hazardous Materials associated with the Property.

The term 'Hazardous Materials' includes, without limitation, (a) material that is flammable, explosive or radioactive; (b) petroleum products, including by-products and fractions thereof; and (c) hazardous materials, hazardous wastes, hazardous or toxic substances, or related materials defined in CERCLA, the Resource Conservation and Recovery Act of 1976 (42 U.S.C. §6901, *et seq*.; hereinafter, 'RCRA'); the Hazardous Materials Transportation Act (49 U.S.C. §6901, *et seq*.; hereinafter, 'HTA'); the Hazardous Waste Control Law (California Health & Safety Code §25100, *et seq*.; hereinafter, 'HCL'); the Carpenter–Presley–Tanner Hazardous Substance Account Act (California Health & Safety Code §25300, *et seq*.; hereinafter 'HSA'), and in the regulations adopted and publications promulgated pursuant to them, or any other applicable Environmental Laws now in effect or enacted after the date of this Conservation Easement.

The term 'Environmental Laws' includes, without limitation, CERCLA, RCRA, HTA, HCL, HSA and any other federal, state, local or administrative agency statute, ordinance, rule, regulation, order or requirement relating to pollution, protection of human health or safety, the environment or Hazardous Materials. Grantor represents, warrants and covenants to Grantee that activities upon and use of the Property by Grantor, its agents, employees, invitees and contractors will comply with all Environmental Laws.

(j) Warranty. Grantor represents and warrants that Grantor is the sole owner of the Property; there are no outstanding mortgages, liens, encumbrances or other interests in the Property (including, without limitation, mineral interests) which have not been expressly subordinated to this Conservation Easement, and that the Property is not subject to any other conservation easement or interest that is adverse to this Conservation Easement.

(k) Additional Interests. Grantor shall not grant any additional easements, rights of way or other interests in the Property (other than a security interest that is expressly subordinated to this Conservation Easement), nor shall Grantor grant, transfer, abandon or relinquish any water or water right associated with the Property, without first obtaining the written consent of Grantee and USFWS. Grantee or USFWS may withhold such consent in its sole discretion if Grantee determines that the proposed interest or transfer is inconsistent with the purposes of this Conservation Easement or will impair or interfere with the Conservation Values of the Property. This Section 14(k) shall not limit the provisions of Section 2(d) or 3(n), nor prohibit transfer of a fee or leasehold interest in the Property that is subject to this Conservation Easement and complies with Section 11.

(l) Recording. Grantee shall record this Conservation Easement in the Official Records of the County in which the Property is located, and may re-record it at any time as Grantee deems necessary to preserve its rights in this Conservation Easement.

(m) Third-Party Beneficiary. Grantor and Grantee acknowledge that USFWS is a third party beneficiary of this Conservation Easement with the right of access to the Property and the right to enforce all of the obligations of Grantor under this Conservation Easement.

(n) Funding. Funding shall be held in trust or by other means specified in the Management Plan for the perpetual management, maintenance, monitoring and reporting of this conservation easement and the Property in accordance with the Management Plan.

IN WITNESS WHEREOF Grantor has executed this Conservation Easement Deed the day and year first above written.

GRANTOR:

BY: ______________
NAME: ______________________
TITLE: ____________
DATE: _________________________

GRANTEE:

BY: ______________
NAME: ______________________
TITLE: ____________
DATE: _________________________

Appendix IV

Template for a Credit Sales Agreement

This template provides an example of the content, structure and language currently included in a conservation bank's credit sales agreement. A standardized template is used to simplify and streamline the bank establishment and management process.

AGREEMENT FOR SALE OF [NAME OF SPECIES] CREDITS

SERVICE File No. ___________________

This Agreement is entered into this __day of ______, 19_, by and between______ ____________________ (Bank) and ________________ (Project Proponent) as follows:

RECITALS

A. ______________ has developed the [Bank Name] located in _________ ________ County, California; and

B. [If applicable] The Bank has developed the [Bank Name] which was approved by [name agencies that approved] on [date approved]; and

C. The Bank has received approval from the US Fish and Wildlife Service (Service) to offer ______________ [type of credits, e.g. vernal pool preservation] credits for sale as compensation for the loss of [species and/or habitat type for which the bank was set up, e.g., federally listed vernal pool species as specified in the Conservation Bank Project Agreement]; and

D. Project Proponent is seeking to implement the project described on Exhibit AA@ attached hereto (Project), which would unavoidably and adversely impact _____________ [species to be impacted by the project, e.g. listed vernal pool crustaceans], and seeks to compensate for the loss of _______________ [species/habitat affected, e.g. vernal pool habitat] by purchasing compensatory credits from Bank; and

E. Project Proponent has been authorized by the Service, Service File No. _____________, to purchase from the Bank _____ [number of credits] _________ [credit type] credits; and

F. Project Proponent desires to purchase from Bank and Bank desires to sell to Project Proponent _____ [number of credits] _____________ [credit type] credits;

NOW, THEREFORE, THE PARTIES AGREE AS FOLLOWS:

1. Bank hereby sells to Project Proponent and Project Proponent hereby purchases from Bank _____ [number of credits] _____________ [credit type] credits for the purchase price of _______________. The Bank will then deliver to Project Proponent an executed Bill of Sale in the manner and form as attached hereto and marked Exhibit AA@. The purchase price for said credits shall be paid by cashier's check or, at the option of Bank, wire transfer of funds according to written instructions by Bank to Project Proponent.
2. The sales and transfer herein is not intended as a sale or transfer to Project Proponent of a security, license, lease, easement, or possessory or non-possessory interest in real property, nor the granting of any interest of the foregoing.
3. Project Proponent shall have no obligation whatsoever by reason of the purchase of the compensatory credits, to support, pay for, monitor, report on, sustain, continue in perpetuity, or otherwise be obligated or liable for the success or continued expense or maintenance in perpetuity of the credits sold, or the Bank. Pursuant to [name of Bank agreement], Bank shall monitor and make reports to the appropriate agency or agencies on the status of any compensatory credits sold to Project Proponent. Bank shall be fully and completely responsible for satisfying any and all conditions placed on the Bank or the compensatory credits, by all state or federal jurisdictional agencies.
4. The compensatory credits sold and transferred to Project Proponent shall be non-transferable and non-assignable, and shall not be used as compensatory mitigation for any other Project or purpose, except as set forth herein.
5. Project Proponent must exercise his/her/its right to purchase within 30 days of the date of this Agreement. After the 30 day period this Agreement will be considered null and void.
6. Upon purchase of [describe credits] specified in paragraph D above, the Bank shall complete the payment receipt form attached hereto as Exhibit AA@, and shall submit the completed payment receipt to the Service.

IN WITNESS WHEREOF, the parties have executed this Agreement the day and year first above written.

BANK
[BANK NAME]

By: ____
PROJECT PROPONENT

[NAME OF PROJECT PROPONENT/COMPANY]

By: ____

APPROVED

US FISH AND WILDLIFE SERVICE:

This Agreement fulfills the ____________________ mitigation requirement, as specified under Service File No. __________ [1-1-__-__-__] dated ____________ ___, 19__.

UNITED STATES DEPARTMENT OF THE INTERIOR
FISH AND WILDLIFE SERVICE

By: ______________________________________

Title: ______________________________________

Dated: ______________________________________

Exhibit AA@

DESCRIPTION OF PROJECT TO BE MITIGATED

[Name of Project (Service File No. ____________)], ____________ County, California

*****

Exhibit AA@

BILL OF SALE

Contract # _______________ [Bank Sales Number]

Service File ___________ [1-1-__-__-__]

[Other Agency file number]

In consideration of $_________________, receipt of which is hereby acknowledged, __________________ [Bank Name] does hereby bargain, will and transfer to ________________ [Project Proponent], _________ credits in the [Bank Name] in ________________ County, California, developed, and approved by the US Fish and Wildlife Service [and any other agencies].

____________ [Bank Sponsor] represents and warrants that it has good title to the credits, has good right to sell the same, and that they are free and clear of all claims, liens or encumbrances.

____________ [Bank Sponsor] covenants and agrees with the buyer to warrant and defend the sale of the credits hereinbefore described against all and every person and persons whomsoever lawfully claiming or to claim the same.

DATED: ____

[Bank Name/Bank Sponsor]

By: ____

Exhibit AA@

[CREDIT TYPE] CREDITS: PAYMENT RECEIPT

PARTICIPANT INFORMATION

Name: ____
Address: ____
____

Telephone: ____
Contact: ____

PROJECT INFORMATION

Project Description: ________________________________
Service File Number: ________________________________
Species/Habitat Affected: ____________________________
Credits to be Purchased: _____________________________
Payment Amount: ___________________________________
Project Location: ___________________________________
County/Address: ____________________________________

PAYMENT INFORMATION

Payee: ____
Payer: ____
Amount: ____
Method of payment:
Cash ___ Check No. __________ Money Order No. _________
Received by: ______________________________
(Signature)
Date: _______________
Name: _____________________________
Title: ______________________

Index

Page numbers in *italics* refer to figures, tables and boxes